Roping in the History of BRONCOING

First published in 2007 by Central Queensland University Press

Second published in 2011 by Boolarong Press, Salisbury, Brisbane, Australia.

National Library of Australia Cataloguing-in-Publication entry

Author:	Lewis, D. (Darrell)
Title:	Roping in the history of broncoing / Darrell Lewis.
ISBN:	9781921920240 (pbk.)
Subjects:	Cattle--Handling--Australia.
	Calf roping--Australia--History.
Dewey Number:	636.2083

Printed and bound by Watson Ferguson & Company, Salisbury, Brisbane, Australia.

in the History of BRONCOING

by Darrell Lewis

FOREWORD

I feel privileged to be asked to write a foreword to Darrell Lewis' book, *Roping in the History of Broncoing*. Years of meticulous research has gone into the writing of this timely volume and for anyone interested in authentic outback history, and bronco branding in particular, this book is in the 'must read' category. The reader will continually find amazing and absorbing information, from the origins of the bronco method of branding cattle up to the present bronco branding competitions.

In the early days on large cattle runs in Northern and Central Australia there were few fences or yards, and no mechanisation, and animals to be branded ranged from calves up to full-grown cows and bulls. The herds often were huge and wild, and the conditions of the time gave rise to some of the country's finest and most capable horsemen. In many areas rainfall was very low, and over long distances there often was very little timber suitable for yard building, so a method had to be found for branding cattle without the benefit of yards. Necessity being the mother of invention, the bronco method for branding cattle evolved.

I can still remember my first effort in the bronco yard, back in about 1937 when I was about twelve years old. The yard was on a claypan just south of Palparara homestead, west of Windorah. The head stockman was Angus Scobie and my job was to keep the brands hot. Over the years I progressed to branding and then to applying the leg ropes, earmarking and using the knife. Finally I made it onto the bronco horse. The early 1950s found me contract mustering and droving for Jack Clanchy at Kamaran Downs, near Bedourie. At that time there wasn't a yard on the place so it was mostly open bronco work. To be done successfully open bronco requires a little more effort and skill, holding the cattle on camp, and at the forked tree.

The bronco method remained the mainstay of cattle-branding for decades, but with the invention of steel portable yards in the 1960s and the introduction in the 1970s of the compulsory BTEC Program – the Brucelosis and Tuberculosis Eradication Campaign – the bronco method started to die out. Many experienced bronco men left the industry or retired, and the younger ringers never had the opportunity to learn the old broncoing skills. Out of concern that the old skills of greenhide rope-making, head-roping and leg-roping would be lost forever, about twenty years ago the bronco method was transformed into a competitive sport, with R.M. Williams being one of the original sponsors. I still had bronco horse and gear in use in 1991 when competition bronco branding began at Stonehenge, Longreach and Winton, so I began entering the various events. Today there is a National Circuit with towns from the Cooper to Camooweal and beyond to the Territory and northern South Australia taking part. The prize money is large, and increasing numbers of tourists attend the competitions to enjoy this genuine Australian outback sport.

I still enjoy taking part in the competition circuit and for old-timers like myself it's encouraging to see younger generations of station workers taking part in these events, with many becoming quite skilled as competitors. In fact, it's not unusual to see entire families involved, with children as young as ten years 'running the brands' while their parents catch on the bronco horse or handle the leg-ropes. The baton has been passed on and the survival of the varied skills of the bronco branding method is assured. I am sure that this book will be an important linchpin in the continuation of this very 'true-blue' bush craft.

Charlie Rayment,

Eildon Park station

Western Queensland

ACKNOWLEDGEMENTS

I began serious research into the origin and history of broncoing in 1997 and since that time many people have contributed with photos, documents and memories. Among the current and former cattlemen and cattlewomen who have assisted are Mick Bower, Katherine, NT, former ringer on Birrindudu, Gordon Downs, Nicholson, Manbulloo and Nutwood Downs stations; Buck Buchester, Wave Hill station, NT, former drover and former ringer on Victoria River Downs and Wave Hill, and one of the great bronco men of the north; Charlie Rayment, Eildon Park station, Winton, Qld, former ringer in western Queensland, and expert bronco man; Lester Cain, Swanvale station, Stonehenge, Qld, former ringer in the Northern Territory and Western Queensland, and expert bronco man; Laurie Bain, cattleman, South Tibraden, WA; Maureen Wood, cattlewoman, Moonyoonooka, WA; Lloyd Fogarty, Timber Creek, NT, former ringer and manager on Auvergne station, NT; Cec Watts, Rockhampton, Qld, former ringer on and manager of Vesteys stations in the Northern Territory and Kimberley; Paddy Cannon, Atherton, Qld; Colin Campbell, Mitchell, Qld; Reg Durack (deceased), son of Kimberley and Northern Territory pioneer M.P. Durack, and former owner of Bullita, Spirit Hill and Kildurk stations; Norm Forster, former ringer in western Queensland, former station manager in the Northern Territory and now station manager in Queensland; Joyce Galvin, Mt Isa, Qld, daughter of the late Dick Scobie, former owner of Hidden Valley station in the Northern Territory; Glen Galvin, Mt Isa; John Graham (deceased), son of Tom Graham, manager of Victoria River Downs from 1919 to 1926, and ringer on Wave Hill station in 1926; Andrew Glenn, manager of Ashburton station, Western Australia; Edie Hackman (deceased), Wandal, Qld, former ringer on Northern Territory stations and later expert sculptor in bronze; Darryl Hill, Katherine, NT, former ringer and station manager in the Victoria River district, NT; Randall Crozier, manager of Anna Creek station, South Australia; John Nicolson (deceased), former ringer on Coolibah, Bradshaw and Bullo River stations, later part-owner of Bullo River station and gold miner at Halls Creek; Ernie Rayner, Atherton, Qld, former ringer on Coolibah, Victoria River Downs and Willeroo stations, former Stock Inspector in the Victoria River District and former station owner in the Northern Territory; Jack Sammon, Rydal, NSW, former ringer and drover in Queensland and the Northern Territory, and bush poet; Charlie Schultz (deceased), owner of Humbert River station, NT, from 1928 to

1971, a great horseman and sometime drover; RM Williams (deceased), former drover, ringer, miner and founder of the famous RM Williams Company; Marie Mahood, Cattle Camp station, Nebo, Qld, pioneer of Mongrel Downs, NT, and now well-known author; Bob Mahood, owner-manager of Cattle Camp station; Stan Jones (deceased), former manager of Gordon Downs, NT; Alan Fennel, manager of Lambina station, northern South Australia; Rodney Watson, Mt Isa, Qld, former Northern Territory and Queensland ringer and drover; Geoff Allen, Wahroonga, NSW, former ringer on Northern Territory and Kimberley stations; Rob Sampson, Moree, NSW, former ringer on Wave Hill in the Northern Territory; Gavin Hoad, manager of Wave Hill station; Des Stenhouse, Katherine, NT, former ringer in south-west Queensland and manager of Northern Territory stations; Ted Fogarty (deceased), Alice Springs, former ringer and station manager, and owner of stations in Central Australia.

I am indebted to the following family, amateur and professional historians: Rans Baker, Wyoming, USA, researcher at the Carbon County Museum; FW Braid, Ballina, NSW, circus historian; Jim Fogarty, Penrith, NSW circus historian; Marion Beattie, Melbourne, Victoria, granddaughter of 'Bronco George' Mellor; Jill Bowen, Forrestville, NSW, former editor of the Stockman's Hall of Fame paper; Steven Butler, Texas, member of web site 'texian1846@yahoo.com', dedicated to nineteenth century Irish author Mayne Reid; Jan Cruickshank, Port Macquarie, NSW, granddaughter of Bob Watson, manager of Victoria River Downs from 1896 to 1900; Deane Fergie, anthropologist at the University of Adelaide; Graeme Gillam, local historian at Warwick, Qld; Philip Gee, William Creek, SA, an expert camel man with a deep knowledge of the Lake Eyre country; Nick Gill, lecturer at the School of Earth and Environmental Science, Wollongong University; Jenny Hicks, Wahroonga, NSW, film-maker and author of *Cowboys, Roughriders and Rodeos*; Chris Hughes, Adelaide, whose forebears were pioneers of Nockatunga station, Qld; Vern O'Brien, Darwin, former Director of the Northern Territory Lands Department and walking encyclopaedia on Northern Territory history; Stuart Duncan at the Office of the Placenames Committee, Lands Department, Darwin; Tony Roberts, Bega, NSW, authority on the history of the Northern Territory Gulf country and author of *Frontier Justice*; John Skinner, Warwick, Qld, former General Secretary of the Australian Rough Riders Association; Bruce Strong, Darwin, Northern Territory historian; Howard Pearce, author of *Homesteads of the Stony Desert*, Brisbane, Qld; Helen Tolcher, Adelaide, SA, historian of north-east South Australia and the Birdsville area of south-west Queensland; Oliver Knox, younger brother of Jack Knox (deceased) who was a ringer on VRD in the 1930s and an expert greenhide rope-maker.

Librarians Jim Andrighetti at the Mitchell Library, Valerie Sitters at the Mortlock Library, and Wendy Tabrett at the Stockman's Hall of Fame generously provided their time to seek out records and photos.

Others who contributed in various ways include Laurie Lewis, Rawson, Victoria; Eric Rolls, Narrabri, author of numerous books, including several wonderful works on environmental history; David Nash, linguist, Canberra; Julia Robinson, researcher at the Australian National Dictionary Centre, ANU, Canberra; Deborah Rose, Senior

Fellow at the Centre for Resource and Environmental Studies, ANU and Lindsay Ward, Stock Inspector, Halls Creek, WA.

My sincere thanks to all these people, and to any who should have been included but have somehow been forgotten. And my thanks to Franz Ranacher, of Bullo River station, who had absolutely nothing to do with this study.

Finally, I gratefully acknowledge the financial support of the Hawke Institute, Adelaide, for publication of this book.

CONVERSION TABLE

1 mile	1.6 kilometres
1 foot	30.5 cm
1 inch	25.4 mm
1 yard	91.44 cm
1 square mile	2.5 square kilometres
1 square mile	259 hectares
1 pint	.568 litres
1 gallon	4.5 litres
1 lb	.45 kg
1 centimetre	.39 inches
1 metre	39.3 inches
1 hectare	2.47 acres
1 kilometre	.62 miles
1 square kilometre	100 hectares
1 square kilometre	.386 square miles
1 litre	.219 gallons
1 kilogram	2.2 lbs
1 pound/'quid' (£)	$2.00 (in 1966)

INTRODUCTION

For most of its white settlement history Australia has been, and remains, one of the great cattle lands of the world. More than a third of the continent – a million square miles – is devoted to cattle raising, and over twenty-four million head now roam its pastures.[1] For decades the settler frontier was a cattle frontier, with great herds being pushed ever on until the last corners of the continent were reached, and today the mythology of the outback is filled with images of the cattle trade – black and white stockmen, the drover and his 'boy', cattle duffers, Aboriginal cattle-spearers and giant cattle stations. To run cattle means to handle cattle – for branding, dehorning, castration and other treatment – and different methods were developed according to prevailing conditions. Eventually one particular way of handling cattle came into general use across the outback – the method known as broncoing.[2]

In simple terms, the broncoing technique involved the roping of cattle from horseback so that they could be dragged to a tree or a specially constructed panel, immobilised, thrown and branded, but there was much more to it than this. The technique is unique to Australia, and quite different from the way American cowboys and Mexican vaqueros used their lassos to handle livestock, well-known from innumerable Hollywood movies. How broncoing was done and the equipment needed is described in detail below.

There are thousands of men and women in Australia today who have 'broncoed' cattle, but few, if any, know anything about the origin of the technique, and in recent times this question has become a matter of some debate. So how did this technique come into being? Exactly how did it work, and how were cattle handled before broncoing was developed? Who developed it, when and where? This book provides answers to these questions. It begins with the cattle-handling methods brought to Australia by the British settlers and then traces the different methods that came into use here. It's a journey that leads into little-known areas of Australian history, and introduces us to a cavalcade of wonderful characters – wild bushmen, innovative cattlemen, Australians ranching in Mexico and America, goldminers, Wild West Shows, cowboys, vaqueros and stockmen. It was the interaction of these characters, played out against the vastness of the outback in times beyond human memory, which eventually gave rise to a specialised cattle-handling technique found nowhere else in the world.

CONTENTS

Chapter 1

CATTLE-HANDLING METHODS IN 17th AND 18th CENTURY BRITAIN

The early settlers in Australia, whether convict or free, were predominantly Anglo-Celtic, and there can be little doubt that in the beginning they raised their cattle more or less as had been done for centuries in England, Wales, Scotland and Ireland. So how were cattle raised and handled in Britain up to the eighteenth century? In two ways it seems, depending on which part of the country the cattle were in. In England the emphasis in land use was on cultivation, so to protect the crops cattle and other livestock were kept in enclosures. Cattle manure was carried to the fields to fertilise the soil, and the cattle were quiet and easily driven on foot.[1]

In Scotland, Ireland and Wales, the method was completely different. There the emphasis was on open range cattle raising and minimal cultivation,[2] and as a result the highland cattle were somewhat wilder than those of the lowlands, although still relatively quiet (a similar open range system was to become the dominant way that cattle were raised in much of Australia for more than a hundred years).

Cattle from the Welsh and Scottish highlands were being driven to lowland English markets by the fourteenth century.[3] This cattle trade increased in importance in the 1500s and 1600s, and there were professional drovers by 1500.[4] The numbers of cattle involved were high, even by later Australian standards. By 1620 Ireland was sending 100,000 cattle to England annually,[5] and in 1662 320,000 cattle passed through the English town of Carlisle.[6] One authority describes the Scottish drovers of the mid-eighteenth century as,

> *great stalwart hirsute men, shaggy, unkempt and wild. They liked their drink, for it was thirsty work walking or riding in the heat of the sun and in a cloud of dust raised by the plodding beasts. Often they would drink too much and annoy the disapproving countryfolk.*[7]

This doesn't sound too different from the stereotypical image of the Australian drover, perhaps, but the same authority says that, 'The drovers often knitted stockings as they walked behind their beasts'![8] This habit was not as strange as it may sound to modern ears; the knitting was being done so that the product could be sold at markets, a means of improving the drover's income which otherwise was not very large.

In the sixteenth, seventeenth and eighteenth centuries many of the drovers went armed because their routes sometimes passed through relatively empty regions frequented by outlaws and armed gangs who tried to steal the cattle.[9] In fact, so much cattle stealing went on in Scotland that one researcher has remarked, 'It almost seems that cattle raiding was in the sixteenth and early seventeenth centuries the chief occupation of the people of Scotland.'[10] Droving was considered of such importance that during periods when the carrying of arms was banned, drovers were exempt.[11] By 1650 the cattle trade had given rise to 'individualistic, privileged entrepreneurs, who pursued a productive capitalistic system of lean-beef cattle raising using hired labour on large, privately owned estates'.[12] These men sound suspiciously like the 'cattle kings' who later would loom large in the history of Australia.

The wealthier drovers sometimes were mounted on ponies but the cattle were usually quiet enough that horses were not necessary, particularly once the relatively narrow confines and ordered fields and laneways of England were reached. Instead, dogs were used and the term 'bulldogging', referring to dogs hanging from the lip or loose skin of the neck of a beast, was in use there by the 1700s.[13]

By the late 1700s there were recognised stock routes to lowland fattening areas and distant urban markets. Some of the lowland roads were stony and made the cattle footsore, so the cattle were often shod,[14] a practice that continued in Australia until the late twentieth century.[15] To shoe the beasts they were first thrown on their backs and had their feet tied together. In Wales the throwing was done by exceptionally strong men who twisted the animal's head in such a way that it fell on its right side, often with such force that the right horn was embedded in the ground,[16] or was seriously damaged.[17]

From at least 1600 cattle were branded, and earmarking and castration were practiced.[18] Unfortunately the available literature does not say anything about how cattle were handled for such treatment, nor is there any mention of specialized stockyards, and the chances are that cattle thrown down in much the same as was done for shoeing.

Chapter 2

AUSTRALIAN CATTLE-HANDLING TECHNIQUES

The First Hundred Years

The first cattle in Australia were large-bodied African beasts with wide-spreading horns and humps, picked up at Cape Town and bought here by the First Fleet in 1788.[1] They did remarkably well. Within twelve months the bulls and four of the cows escaped into the bush. By 1796 they'd increased to 'upwards of forty', by 1800 to more than 300 and by 1810 about 5000.[2] As well as this natural increase, within a decade the founding stock had been supplemented with cattle from diverse sources – more from Africa, zebus from India, one from California, one from Spain, and a number of English shorthorns.[3] By 1820 there were more than 100,000 cattle in New South Wales[4] and by the time the last corners of the continent were settled, a century after the first cattle arrived, the founding herd had increased to 15,000,000.[5]

There can be little doubt that during the first decade or so of settlement cattle were handled more or less as they had been in the mother country, but as country 'further out' was taken up, methods changed. An account of the colony published in 1819 said that in the region around Port Jackson, 'horned cattle are followed by a herdsman during the day...and they are confined during the night in yards or paddocks' to prevent them from wandering onto croplands, but 'In the remoter districts...which are altogether devoid of cultivation, horned cattle are subjected to no such restraints, but are permitted to range about the country at all times.'[6]

Running cattle on the open range with minimal handling by stockmen quickly led to the cattle becoming very wild. In 1827 Cunningham described,

> *the calves becoming as wild as deer and almost as fleet too, the herd in fact requiring to be hunted into the stock-yards by bands of horsemen, and thence culled as required. A stock-yard under six feet high, will be leaped by some of the kangaroos (as we term them) with the most perfect ease, and it requires to be a stout as it is high to resist their rushes against it.*[7]

Cunningham went on to describe how, 'When an ox is wanted for killing or branding, a noose is thrown over its horns, and the rope carried around a post, whereto it is dragged.'

As new and remoter country was settled and the size of the landholdings increased, the herds grew ever larger and wilder, and new methods had to be devised to muster, brand, castrate, dehorn and in other ways treat cattle. A good coverage of early Australian cattle-handling techniques has been put together by John Perkins and Jack Thompson,[8] and this shows that comparatively little is known about how cattle were handled until the 1830s. This can be at least partly explained by the relatively low numbers of cattle in the colony during the first two to three decades of settlement. My own research confirms the scarcity of pre-1830s accounts although from later times there are a few descriptions of branding methods not listed by Perkins and Thompson, including the broncoing technique. More than one method was used at any one time and sometimes two were used at the one place. Until the late 1800s almost every method required the use of a yard.

When it came to calves probably all methods involved separating them from their mothers and 'scruffing' them (throwing them down physically; see plate 1), but large wild cattle were a different matter. One way to handle such animals was to use a roping pole, a long pole with a noose draped over the end which enabled a man to place it on a beast from a distance (plates 2, 3 and 4). This was being used in New South Wales and Queensland by the 1840s[9] and its use continued to the early 1890s,[10] and even, on occasion, as recently as the 1930s.[11] Clumsy and slow as it undoubtedly was, the roping pole could be applied without too much trouble to staid cattle or to wild cattle packed into a yard, but it could be dangerous when applied to a plunging horse. Writing in 1951 Arthur Campbell, presumably an old man at the time, said that,

> *In my youth I remember watching old hands making frantic endeavours to place a rope that was on a pole over a colt's head. By this method they sometimes succeeded, but on other occasions they knocked out the colt's eye in the process. The rope, once on, we were instructed amidst much profanity to take a turn around the corner post, while the breaker, with a halter on a stick, tried to put it on the colt's head. The victim usually choked down bereft of breath, and at times of life itself.*[12]

Once caught with the roping pole, men dragged the beast to a place where it could be immobilised for branding and other treatment. A description of a branding system used on the Monaro Tableland in New South Wales in the 1840s relates how,

> *One man, the best that can be found, then goes into the yard with a noose rope on the end of a long forked pole, the plain part of the rope running through his hand and trailing along the ground; the end, however, if long enough, being held outside the yard. This noose is then dexterously thrown over the beast's horns or round his neck, and a turn being taken round one of the round corner posts at the yard, the herd is driven toward that corner, and the slack of the rope taken in till*

> *the beast that is noosed is drawn taught [sic] up to the post. Then sometimes the animal is branded standing, his legs merely being confined by a leg rope held by men or fastened to another post; or else, if very wild and powerful, he is legged, and thrown, and tied fast, and then branded.*[13]

As well being dragged to a corner post, cattle could be hauled into a 'sort of wooden cage' (plate 3)[14] or to the side of the yard (plate 5),[15] or to a 'branding panel',[16] a free-standing set of strong posts and rails close to and parallel with the side of the yard. The only use mentioned for this panel was for the protection of men being charged by an angry beast, but the name suggests that it also had roped cattle dragged up to it.

Another branding method was being used in Queensland, New South Wales and possibly Victoria in the 1840s, 1850s and early 1860s. A woman who travelled to New South Wales from Western Australia in 1858 was surprised to see large herds of cattle being branded 'without roping them!' She described how they were driven 'into a long wooden yard over which there is a platform' from which they were branded from above.[17] This method is the subject of a painting made in Victoria by S.T. Gill some time before 1865. It shows a mob of cattle closely packed into a solidly made yard that has three large rails between each post (plate 6).[18] In one corner two planks have been laid across the top of the middle rails to form a platform about the height of the backs of the cattle. A man standing on this platform is being handed a branding iron about two metres long which he undoubtedly is about to use to brand the cattle from above. In another corner of the yard a man wielding a stick appears to be trying to force the cattle to move about so that unbranded cattle will come within reach of the man with the branding iron. A painting by George Fairholme shows the same method being used at Canning Downs (near Warwick, Queensland) in the early 1840s (plate 7).[19]

Some early illustrations show cattle yards up to three metres tall,[20] a height perhaps more appropriate for yarding kangaroos than cattle, but possibly made this way to facilitate branding from above, and to circumvent the amazing ability of some wild cattle to leap out of yards (plates 3, 8). On the latter point, a settler in the 1840s had this to say:

> *When some of them get up [from being branded etc.], everybody must be out of the yard. Some degree of nerve is required to untie a beast; the best way is to keep behind it, and out of the way. Some men will stand their ground, and some always nip up over the fence as speedily as they can. Generally the danger is more in appearance than in reality. Only now and then, when a real "Russian" happens to be among the mob, circumspection must positively be practised as well as bravery. I have known beasts to break three strong ropes one after the other, charge everybody out of the yard, and then go over a six-rail fence at a flying leap, and get away unconquered to their wilds again.*[21]

In the 1850s in southern Victoria Cuthbert Fetherstonhaugh saw a man using an unusual method to catch individual animals for branding and ear marking:

> *Out from Portland for about twenty-two miles the road runs through what was all looked upon as barren heath country, though some of it was heavily timbered... There were lots of wild cattle on that heath country in 1854. I knew a man who got together quite a good sized herd there. He used to ride up to a "clear skin," [sic] catch it by the tail, throw it, tie it up, make a fire, brand it, and ear-mark it. (He carried a brand on his saddle.) I have never thrown a beast in that way myself, but I have seen it done. The beast goes over quite easily, but you must be going a fair pace.*[22]

This method did not require a yard, but was of no use when large numbers of cattle needed to be branded. At the time it appears to have been a method of catching and immobilising individual beasts that was idiosyncratic, or used by very few, and it does not appear to have ever come into more than occasional use anywhere in Australia. It clearly was different from the bull-catching method that came into widespread use in outback Australia in the 1960s which required the horseman to dismount before throwing the bull by the tail.

By the 1870s in Queensland another method was in use for branding (plate 9). The following description comes from Gracemere station, near Rockhampton:

> *When cows and calves are 'cut out' they are driven into the stockyard at the head station. Here the calves are drafted apart from the cows, and are then driven one by one up a narrow passage at the end of which is fixed a sort of trap in which each is caught by the neck. Ropes are then slipped over his legs, he is thrown on his side and branded, ear-marked, etc., upon which he is released, the whole operation lasting something under a minute for each calf.*[23]

'Cut out' refers, of course, to mounted horsemen separating a particular class of cattle from a herd being held in the open by other mounted horsemen, a method that's been used in Australia since at least the 1850s (plates 10, 11).[24] On one memorable occasion on the plains of north-west New South Wales in about 1860, a mob of 20,000 head was mustered and close to one hundred men worked to cut out different classes of cattle.[25]

Yet another method involved men on foot roping cattle in a yard and a horse outside the yard being used to drag the roped beast to the rails. The horse being used this way was fitted with the harness normally used to haul a wagon or plough. The rope was attached to a swingle-bar behind the horse which was then walked backwards and forwards as required, and from constant repetition both the man and the horse knew exactly where to stop.[26] In this way the roped beast was dragged up against the side of a yard where it was leg-roped, thrown, branded and otherwise operated on. When the roped animal was immobilised the rope was released from the swingle bar and the man doing the roping pulled enough of it back into the yard to be able to rope another beast. One man who used this method claimed that even though 'it was done over and over again, there was no monotony about it,'[27] but another recalled, 'What a dreary job this was, leading the calf horse, and we all tried to dodge it. Walking the horse out twenty-five to thirty feet, releasing the rope and then back again.'[28]

An account from 1892 describes how, 'Cows with young calves have to be taken quietly; if not, many of the young calves will get fagged ("knock up") and have to be left behind, thus necessarily increasing the number of calves that will be old for branding next time'. The same account goes on to say that,

> *When yarded for branding the calves will have to be drafted from their dams through a lane, the cows generally running off in small lots, while the calves are frightened back by the gesticulations of the man drafting...When the calves are all together [in a yard separate from their mothers]...it is well to go through them first, branding all such as are so old or large as to require roping... Calves not over three or four months, two active men can readily "scruff."*[29]

While the above account does not state where this method was used or explain how the roped calves were dragged to a position for branding, it appears to assume that the use of a yard was normal practice and makes no mention of any method that did not require a yard.

Branding on the Great Inland Plains

Most of the accounts of the different cattle-handling methods described above come from coastal and near-coastal regions of eastern and south-eastern Australia where, relatively speaking, land was fertile, water supplies and seasons reliable, timber for yard building abundant, and properties small. In New South Wales the western plains were settled by the 1840s, and in central-eastern South Australia and south-west and western Queensland they were settled by the 1860s.[30] By the early 1870s stations had been established right out to the Queensland-Northern Territory border and up to the Gulf, and through northern South Australia into Central Australia.[31]

In these regions the land was semi-arid, seasons highly variable, water supplies unreliable and the carrying capacity of the land was light. To compensate for these disadvantages stations needed to be much larger, and this added to the difficulties of managing cattle using the old methods. Descriptions of the techniques being used on the great open plains of the interior in this period are rare. Nevertheless, there is evidence that methods similar to those used closer to the coast, or at least, methods in which a yard was absolutely essential, continued on most outback stations until the end of the nineteenth century, and in some areas well into the twentieth century.

For example, the system that required men on foot to rope a beast which was then dragged to the rails by a horse being led back and forth outside the yard was still in use on Sturt Creek station (south-east Kimberley) in 1902 (plate 12).[32] The same method was still being used on Augustus Downs in the Gulf country in the early 1920s,[33] although by this time the method was long out of date. Photographs from Victoria River Downs (VRD) in the period 1910-1920 show cattle in a yard being roped by men on foot, while outside a mounted horseman waits to drag the roped beast to the side of the yard (plate 13). While this may be a variation on the method used on Sturt Creek and Augustus Downs, it's more likely that the bronco method was being used on VRD at this time, and that this yard was too small for the horseman to work in.

The major disadvantage of the old systems was that they required the use of yards, and even where suitable timber was close at hand yards were costly to build. Given the vast size of the inland cattle stations, the lack of suitable timber in many areas and the higher wages for stockmen, yards were even more expensive and it was economically impracticable to build them everywhere they were needed. An extreme example of the problem comes from the vast open grasslands of Lake Nash station on the Northern Territory-Queensland border. There in 1914 suitable timber for various jobs had to be cut in Queensland and then carted 250 kilometres to the station.[34]

For years after they were formed many inland stations had few yards, and even in the late 1920s some stations still drove their cattle to yards for branding, as the following quote written in 1931 attests:

> *Some cattlemen drive all cattle to one big central yard, some having to be driven forty to fifty miles. The disadvantages are obvious, as it takes days for the cattle to get back to their camps, and calves are split off their mothers and die. The extra handling has the effect of quietening the cattle. Two or three small branding yards, supplementing the main set of yards, are most useful, and do not cost much. All that is necessary are the receiving and forcing yards, lane, pound, calf pen, and spare yard. If thought necessary, a crush and dip can be added, but without them the set can be erected for £100, if timber is handy and good.*[35]

As the above passage indicates, in the absence of yards in many areas unbranded cattle, sometimes weakened by drought, had to be mustered and driven to the nearest yard, often several days or even a week away.[36] This slowed down the branding process and stressed the cattle and 'knocked them about',[37] and increased overall expenses. Sometimes the lack of feed and/or water near the yard would make it impossible to bring cattle there and there was no choice but to leave them until conditions in the vicinity of the yard improved. By the time this happened they might be fully-grown and much harder to handle, or perhaps have been branded by someone else. In spite of these disadvantages, for thirty, forty and even fifty years, the settlers on the inland plains continued to use the old techniques requiring the use of a yard. The new broncoing technique overcame this problem.

Chapter 3

OPEN BRONCO

Broncoing began as a way of handling cattle in the open without the benefit of a yard, a technique that became known as 'open broncoing' (plates 14, 15, 16). To apply this method cattle were mustered and taken to an area of open ground near some trees where they were held together by a number of horsemen, in the same manner as on a drafting camp (as opposed to a drafting yard). In a drafting camp cattle are held together by horsemen who, in effect, take the place of a yard. While they are held other men on specially trained horses ride into the herd and 'cut out' a required beast by gently moving it to the face of the camp, and then pushing it quickly out to where other horsemen are waiting to hold it. This was done to separate animals with different brands, branded animals from unbranded, or bullocks to be delivered to a drover, but open bronco work had quite a different aim and required a completely different method for removing each animal from the mob.

When the cattle reached the chosen area one of the trees was prepared for use, either by cutting one branch off the tree to form a hook into which the head rope could be placed as the horseman rode past (plate 14), or by attaching an iron hook to the tree trunk (plate 15). Ideally a tree with a fork beginning at or near the ground was used because a higher fork could cause calves or smaller cattle to be dragged upwards and choked as they neared the tree. The top of the hook or fork had to be at about waist or chest height – low enough for the head rope to easily be placed in it, but high enough to not allow the rope to accidentally come out as the roped beast struggled. After the fork was prepared or a hook attached, two short stakes were cut and set in the ground at appropriate places a few metres from the tree. These were 'leg rope posts'; if other trees of suitable size were close enough they could be used instead. Meanwhile, a fire was lit to heat the branding irons and when the tree and posts were ready and the branding irons were hot enough, the horseman (known as a 'catcher') entered the mob of cattle to rope a beast.

The length of the greenhide ropes used to catch the required beasts varied according to the preference of the rope-maker, or other factors, but for use in a yard they were usually about eight to ten metres long, and longer if they were used on an open camp or to rope bulls and other large animals. Three to four metres would be taken up in the noose, leaving five to seven metres from the noose to where it was attached on the near side of the horse. The rope was attached to the harness just behind the left hand leg of the catcher, and passed outside his leg and from left to right over the neck of the horse where the noose was held in the catcher's right hand. Many catchers tucked some of the excess length of the rope up under their buttocks (on the near-side), to prevent it dragging on the ground and being trampled on, or possibly becoming tangled in the legs of the horse. The rest of the excess was coiled and held in the left hand, along with the reins (plate 17).

As he rode into the mob a right-handed catcher would hold the noose in his raised right hand (cover photo and plate 17). When he found a beast in the right position for roping (ideally facing towards him on the left or near-side) he would swing the noose over the neck of his horse (from the off-side to the near-side) and onto the chosen animal (some of the more expert catchers could flip the noose in the other direction to rope a beast on their off-side). As soon as the noose was over the animal's head the catcher would give the rope a jerk to tighten it (plate 18), and then turn his horse and move towards the fork or hook. As he did this he would hold the rope up in his left hand until the strain came on, to prevent the loose rope from being fouled by other animals in the yard. If he was holding some of the excess rope under his buttocks it was easily released by the pull of the beast or by the rider easing himself forward in the saddle.

Each animal to be branded was roped in turn and dragged from the mob to the tree. As the bronco horse moved past the tree the rider or one of the men on the ground lifted the rope into the fork or hook. When the beast reached the tree, men on foot roped a front and hind leg and wrapped the ropes around the leg rope posts, taking up any slack as opportunity allowed. The preferred side for branding was the near (left) side, so the animal either fell or was pulled onto its 'off' (right) side, held down and the rope released. While the branding and other procedures were carried out the horseman gathered up the rope, returned to the mob and roped another animal. Because it was difficult to hold the reins while roping an animal or coiling the rope, the catchers often would tie the reins together so that they could not accidentally fall to the ground and be trodden on by the horse.

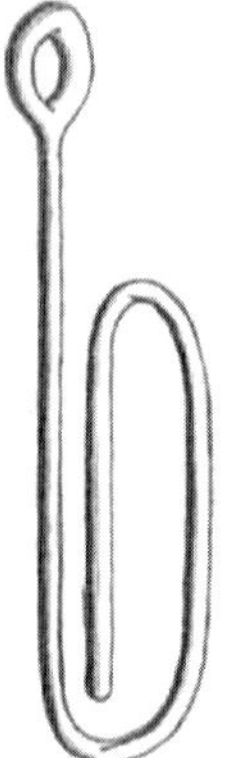

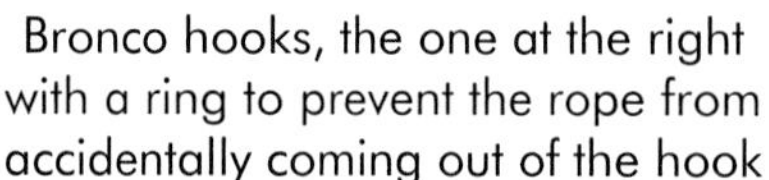

Bronco hooks, the one at the right with a ring to prevent the rope from accidentally coming out of the hook

The iron hooks used could be a simple 'J' shape, but a more efficient design was something like an oversized paperclip with part of the outer curve cut off (see plate 15). This shape meant that if the rope flipped up against the outer side of the hook it would meet the curve at the top and not come out. Some hooks had an eye at the top so that when they were fixed to a tree or post they couldn't swing around, and some had an iron ring attached to the inside end of the hook in such a way that the rope could pass into the hook, but there was no chance it could come out by accident.[1]

Charlie Rayment is a senior cattleman who has worked throughout south-west Queensland and is now the owner of Eildon Park station, south of Winton. He has branded untold thousands of cattle using the open bronco method; one place he helped muster in the early 1950s was Kamaran Downs near Bedourie which had about 10,000 head, but not a single yard on the place.[2] Charlie said he preferred to have the men holding the mob to ring them slowly anti-clockwise so that the catcher could enter the mob and ride clockwise, and thus have cattle to be roped passing on the horse's near-side, the easiest way to catch them. As each calf was released after branding it was pushed around so that it faced the mob because it would then go straight back to its mother, rather than just going bush, and it was good to have a dog to nip the calf to encourage it on its way.[3]

Australian open bronco is superficially similar to the Mexican-American roping method, but there are important differences. While many Mexican vaqueros and American cowboys possessed astonishing skills with the lasso, the inheritance of several centuries of open range cattle raising, the image of a cowboy galloping flat out while twirling a lasso overhead and sending the loop some extraordinary distance to catch a fleeing beast probably has more to do with Hollywood than it does with practical ranch work. Just as was done in Australia, vaqueros and cowboys mustered their cattle and held them quietly on camp while the required beasts were lassoed and dragged out of the mob. However, they were not dragged to a tree or any other sort of fixture to be immobilised. Instead, if the lassoed animal was small men on the ground 'wrestled' ('scruffed') it, and if it was large another horseman lassoed its hind legs, and both horsemen backed up their horses until the roped animal fell over (see plates 19, 20, 21).[4] Another difference is that the American saddle had a horn to which the lasso was either tied fast or looped around so that it could be 'dallied' – rapidly fastened or loosened at will.[5] In Australia the rope used to catch the cattle, called a 'head rope' and made of greenhide, was fixed to a harness arrangement on the side of a specially trained 'bronco horse' or 'bronco mule'.

The Pros and Cons of Open Bronco

From the time it first appeared, opinion has been divided over whether open broncoing was good or bad. Henry Lamond was working on Queensland stations when the broncoing technique first arrived, or shortly afterwards, and years later he recalled that,

> *Arguments about its merits or otherwise were fairly hot and frequently furious... Some cattlemen condemned it on the ground that it made cattle wild... Others took the reverse view. Around camp-fires I've heard old cattlemen state with emphasis and lurid oaths that bronchoing was a primary cause of lumpy jaw. Others were just as vehement lumpy-jaw had nothing to do with bronchoing.*[6]

Concerns about broncoing causing 'lumpy jaw' in cattle were misplaced as it is now known that the swellings and abscesses on the animal's head were symptoms of a chronic inflammatory disease caused by a parasitic fungus.[7]

Some cattlemen regarded open bronco as a 'horse-killing game' because it required constant galloping back and forth to keep the restless herd together. For example, Cec Watts, a long time Vestey Company employee, said that,

> *It was gallop from "go to whoa" as the cattle in the mob reacted to the presence of a roper in their midst and it was hard to hold the mob within reasonable range of the branding panel. Many of the branders arrived at the panel on their sides half choked after the long drag from the mob. Half the stuff that was branded headed bush, as soon as it was let up, ignoring the rest of the cattle. It may well have been an expedient method to hook a few cleanskins out of a small mob but trying to hold a fair mob close to a panel and brand a lot of calves was a very tough task.*[8]

Charlie Schultz, owner of Humbert River station from 1928 to 1971, also considered open bronco work a horse-killing game and said that sometimes after a beast was released it would rush back to the mob and go right through it and out the other side, and that this would stir up the other cattle and make it hard to hold them.[9] Buck Buchester, a former drover and cattleman who worked in the Victoria River district for more than fifty years and who did a tremendous amount of bronco work, agreed with the assessments of both Charlie Schultz and Cec Watts. Buck believed that cattle going bush from the bronco tree or going straight through the mob and out the other side turned wild cattle into rogues.[10] Buck's view was shared by 'Culkah' (J.K. Little) who began work on cattle stations in about 1890 and went on to manage properties in the Northern Territory, Queensland and New South Wales. At various times he was also a drover, a Boer War veteran and commander of a unit of the notorious Queensland Native Police.[11] In 1950 he wrote that 'some of you may have heard of thousands of bulls being shot in one year on some station, and it is not unlikely that lots of those beasts were escapees from an open bronco camp.'[12]

One writer noted that very few calves were parted from the cow when the bronco method was used, but that the bucking and bellowing of the calf as it was dragged from the mob made its mother very anxious. As a result she would often follow her calf and sometimes come too close for the comfort of the men.[13] Jim White, who did a lot of open bronco work on Brunette Downs (Barkly Tableland) between 1928 and 1945, was critical of the method for yet another reason:

> *what a mess there was at times. Rope a young mick, mark and brand him and away he would go, bellowing like a beauty with the mob in hot pursuit. Perhaps the same thing happened a dozen times when it was necessary to almost re-muster the mob over and over again.*[14]

However, another cattleman claimed that beasts rushing back into the mob could make roping the other cattle easier: 'When a calf is let go [after being branded, castrated etc] it rushes back to the mob and the unbranded ones work out to the outside and come easily to hand.'[15]

One of the most vehement critics of broncoing was Jack Kilfoyle, the owner-manager of Rosewood station and in his time one of the most progressive of north Australian cattlemen. Jack's concern was with the effects on the cattle:

> *We always use branding-yards. I would never permit the introduction of the bronco method at Rosewood. I think it contributes to high mortality amongst the cattle and is a contributing factor to injuries and disease as well. It breaks down the constitution on young cattle, some of which never get over the initial damage done by the bronco methods...*[16]

Some men thought the method was all right if you had good horses and enough men, and that it helped 'educate' the cattle. Charlie Rayment acknowledged that bad handling during open broncoing could cause problems later, but he believed if things were done properly cattle were subsequently easier to handle because they were accustomed to being held in a mob.[17] He said that if cattle had never been held in a yard or open-broncoed before, they 'tend to stir up a bit', but 'In the channel country, Birdsville, Bedourie area, where it was done constantly, the cattle were okay to hold on camp, with skilled men.'[18] Charlie added that horses used to hold cattle on an open bronco camp made good night horses because they were trained to watch cattle and push them around and keep them together.[19]

Clearly, the major advantage of the technique and the reason it was invented was that it enabled cattle to be branded and otherwise treated on their home range without having to be moved to a distant fixed yard. Henry Lamond heard many arguments for and against the technique, but he said, 'there was one factor they could not overlook; its economy. It could be done to excess, and prove cheap and nasty. But, taken all in, it was half the answer to the cattleman's dream. The other half hasn't been solved yet. In fact, I doubt it's been invented.'[20] The fact is that whatever its drawbacks, open broncoing caught on and was used for many decades on stations across the outback. Just how many decades is revealed in the following chapters.

Chapter 4

THE ORIGIN OF BRONCOING

The previous chapter describes how the open bronco method worked, but the questions remain as to when and where it began, and who invented it. In recent years various suggestions have been put forward, but most of these have been nothing more than guesses and 'gut feelings'. One old cattleman who wrote to the Stockman's Hall of Fame in 1997 was particularly concerned to deny that 'some Yank came out to Australia and showed our pioneer cattlemen how to brand wild cattle.' Even though no American has ever made such a claim, he warned that 'The Yanks – God bless 'em – will claim anything', and he went on to suggest that '*broncoing* – which was the term we always used out on the stations, not *bronco branding* – was developed ad hoc, out of necessity, probably in the Gulf Country of Queensland, and spread from there.'[1]

Another elderly cattleman refused to accept documentary evidence alone because he couldn't believe that Australian cattlemen wouldn't have invented some way of branding without yards as soon as conditions made it desirable. He believed that early cattlemen probably were roping cattle in the open long before the first written descriptions of the technique appeared.[2] However, the fact is that there are sufficient written descriptions and statements by old-time cattlemen to determine which techniques were being used at different times in the past.

The general ignorance amongst senior cattlemen about the origin of broncoing suggests that the technique was developed in a period beyond human memory and was not well documented at the time; photographic and documentary sources support this contention. Credit for introducing broncoing has been given to a number of people over the years. Foremost among them are a Mexican named Ambrose Madrill and an Australian cattleman named H. 'Compie' Trew, but a cowboy named Joe Case, a Mexican named Corelli, and a Spaniard named Bralla have also been put forward. Clearly the broncoing technique is unlikely to have been developed independently by all these men so in the following each candidate is examined in turn.

Joe Case

Joe Case was an American or Canadian who came to Australia in 1911 and spent the rest of his life working on Willeroo, Waterloo and other stations in the Victoria River district of the Northern Territory.[3] Old time Victoria River cattleman Charlie Schultz knew Case and said that he'd been sent to Australia to inspect properties, and suggested that it might have been Joe who introduced the bronco technique.[4] Another Northern Territory cattleman, Cec Watts, told me that he'd heard that Joe had introduced broncoing to the Kimberleys, and Cec had found the remains of a western saddle on Turner station that was said to have belonged to Case.[5]

Ted Morey, a policeman in the Victoria River district in the early 1930s, saw Case using a lasso.[6] Morey believed that along with three other American cowboys, Case was brought out to Brunette Downs to teach the stockmen American roping techniques for branding in the open, but he says that the technique never caught on. While it is clear that Case used his American roping technique in Australia, he couldn't have been the originator of the bronco method because, as will be seen, it was in use in Australia earlier than his arrival in 1911.

Corelli

In the 1940s Jack Kelly carried out extensive field surveys of the Australian cattle industry for the Federal Government. During one of his surveys he visited Pandi Pandi station in north-east South Australia and there he was told by the station owner, C.C. Morton, that the bronco method was introduced on Pandi Pandi in 1900. According to Morton, before 1900 cattle were handled in big drafting and branding yards. He said that the manager of Pandi Pandi in 1900 was a Mexican named Corelli, and that he roped calves in the open from a Mexican saddle, in the manner of the old-time cowboys on American cattle ranches. Morton said that Corelli's method was subsequently modified to produce the Australian bronco technique.[7] There does not appear to be any contemporary documentation about Corelli, but if Morton was correct that Corelli began using his technique in 1900, there is good evidence that the roping of cattle from horseback was already being practiced in Australia and was introduced by someone else.

Bralla

The only information available about Bralla comes from Helen Tolcher, an authority on the history of Birdsville, the Birdsville Track and the Strezlecki country, and author of a number of books on the region.[8] One of Tolcher's books is a history of Innamincka, and in this she mentions that the Brallas were a Spanish family that came to South Australia with the first shipment of donkeys imported by Thomas Elder.[9] Another of Tolcher's sources states that the original Bralla was a noted horse-breaker.[10]

The donkeys imported by Thomas Elder were a mob of twenty-seven that arrived at Port Augusta from India in January 1866, on the same ship that brought out 120 camels for Elder.[11] If the Brallas were Spaniards it seems unlikely that they just happened to be in India at the right time for the husband to be hired to accompany the donkeys to Australia, so either the story is incorrect or they came out with another as yet undocumented import of donkeys from Mexico or Spain. In any case, the roping skills possessed by Mexican cattlemen did not come from Spain,[12] so unless Bralla had been to Mexico it's unlikely that he demonstrated Mexican-style roping skills here or that he was somehow the originator of the Australian bronco method.

Ambrose Madrill

The claim that Ambrose Madrill introduced the bronco method comes from the book, *Packhorse and Pearling Boat*, written by Tom Ronan.[13] Tom's father, Jim, had worked throughout western and south-west Queensland, north-eastern South Australia and north-western New South Wales in the 1880s and 1890s, including some years as cattle and horse buyer for the 'Cattle King', Sidney Kidman.[14] Jim told his son that Madrill had come to Australia in charge of a mob of Mexican jack donkeys, imported to Blanchewater station for mule breeding by Thomas Elder, and he said that at Blanchewater Madrill,

> *was bewildered at the local stockhandling methods. He had been trained in practices evolved in Mexico during two and half centuries of open-range grazing... Madrill volunteered to demonstrate how such matters were ordered in his homeland. He had a rough panel of posts and rails erected on a patch of clear ground. While some of the hands mustered and held the cattle in a close mob, Madrill, with his own rope hooked to his own Mexican-style saddle, rode in among the stock, lassoing each calf and dragging it up to the branding team at the panel. The new idea became widely accepted.*[15]

There are points for and against Ronan's story. Those against include the fact that the only donkeys known to have been imported by Thomas Elder are the twenty-seven mentioned above that arrived from India. There's no record of Elder importing donkeys from Mexico and taking them to Blanchewater station although it remains possible that he did so. Elder didn't own Blanchewater until 1872,[16] so this is the earliest date that Madrill or anyone else could have worked there for him.

Another point against Ronan's story is that during open branding in Mexico, depending on the size of the animal cattle to be branded they were roped by one or two mounted horsemen, as described in the previous chapter, and there's no evidence of any kind of branding panel (or forked trees) being used there or anywhere else in the Americas. If Madrill ever demonstrated the Mexican open range technique of roping cattle from horseback and dragging them to a branding iron fire, he wouldn't have constructed a panel against which to drag the roped beast.

Points in favour of Ronan's story include the fact that Madrill certainly existed. Exactly why or when Madrill arrived is unknown, but he had a family in Australia; two of his sons ended up in the Northern Territory as station owners and one of them was a packhorse mailman there in the early 1900s.[17] The man Bralla discussed above is also said to have come out with donkeys imported by Elder, so it may be that Bralla and Madrill came out together. Between 1872 and 1890 Blanchewater was used primarily for horse breeding (and probably mule breeding), but in 1891 the horses were sold off and the station was given over to sheep.[18] If Madrill or Bralla ever were on Blanchewater they would have been there between 1872 and 1891.

Being a Mexican Madrill almost certainly brought a Mexican saddle to Australia and showed Australians how it was used, but there's no evidence to substantiate Ronan's claim that he was the originator of the broncoing technique. If he came out with donkeys imported by Elder in 1866 or with an unknown import in the 1870s or 1880s, and introduced the Mexican roping technique soon afterwards, there should be evidence of its appearance in the historical records, but there's no such evidence. What evidence there is suggests that roping from horseback was introduced in 1891 by someone else, and that in many areas the pre-broncoing techniques were being used well into the 1900s.

'Compie' Trew

H. Compton 'Compie' Trew was born in South Australia in 1864.[19] His began work on Mundowdna station near Marree in 1880 and spent his entire working life in cattle country. He had a long life, dying in Adelaide in 1958 at the age of ninety-four,[20] and he would have been familiar with whatever cattle-handling techniques were in vogue from 1880 until he retired, probably in the 1930s (plate 22).[21] The late R.M. Williams knew Compie Trew personally and had the following to say about him:

> *Old Pompey[22] Trew lived most of his life on the Birdsville Track. He had a small block on the north-east side of Lake Eyre. He was a great innovator...a man of new ideas. He first recommended the height and pitch of steel in a [pack]saddle tree. He also recommended (it was his idea) that a breeching and breastplate were unnecessary in a packhorse. He suggested a tree that would take a flank girth well back. We manufactured his idea. When we made the first one Sir Sidney Kidman bought it on the spot... He invented a set of [bronco] harness with a heavy ring (six inches) on the shoulder. We made it up for him [the R.M. Williams Company]. We were the first to make that particular stuff that he invented. In Pompey's earlier days it was done with a collar (easier on the horse if the horse was working for long hours). Give Pompey the credit for the high...saddle tree.[23]*

In his book, *The Drovers*, Keith Willey refers to Trew's 'practical genius' and declares that 'his ideas revolutionised the organization and equipment of pack trains which carried supplies for drovers on the Birdsville Track and to the lonely stations of the desert fringe.'[24] He gives Trew the credit for inventing the 'Bedourie packsaddle' (probably the 'high tree' saddle tree that Williams mentions), and the Bedourie oven. In other words, Trew was an innovator and a man who constantly tried to improve equipment and methods.

A number of sources credit Compie Trew with inventing the bronco system. One is A.M. Duncan Kemp, author of *Our Sandhill Country* and other books, who says that Compie Trew was 'one of the first to introduce the lasso or broncho work'.[25] Duncan Kemp's parents moved to the Birdsville country in the late 1890s and raised their family there,[26] so she was in a position to learn about the origin of broncoing from her parents and other locals. Another source is an article on broncoing written by Alec Marshall and published in *Walkabout* magazine in February 1937. Marshall says that he was using the bronco system in south-west Queensland in 1910 and states that, 'It is a debatable question just how long "broncoing" has been in vogue; but it probably was first introduced at Goyder's Lagoon cattle station [part of Clifton Hills] by the then Manager, a Mr. Trew, some time in the present century.'[27]

A third source is a statement by Artie Rowland, who was born in about 1886 and who was a highly respected cattleman in the Birdsville-Coopers Creek country in the first half of the 20th century.[28] In 1975 when he was in his eighties Rowland was interviewed by Helen Tolcher, and he told her that Trew was the originator of broncoing:

> *Trew, old Compy Trew...he was the first man that started the broncoing. ... Old Conrick at Nappa Merrie, he said, "By Gad, I'm not having my calves choked, run 'em up to trees and that sort of thing." They all started to...the mission people at Koppermana [sic] was the second, the first, I mean, the first to start it, then it spread everywhere.*[29]

Rowland also stated that Clifton Hills and Goyder's Lagoon were run as one station and that Trew based himself at the latter place.

Finally, when Trew died in 1958 an obituary for him had this to say:

> *Mr. H. Compton Trew, a veteran cattleman, and one of the most colourful personalities of the Australian outback, died at his Glenelg home yesterday. He was 94. Mr. Trew was famous throughout the inland for introducing the system of using greenhide like ordinary rope to lasso cattle for branding.*[30]

So, is there any evidence to back up these claims? In 1963 Henry Lamond wrote an article on how to make twisted greenhide ropes[31] and in it he recalled reading about 'bronco branding' in *The Pastoralist' Review* in 1905, but he couldn't remember the name of the author. The author was, in fact, Compie Trew.[32]

Chapter 5

TREW'S LASSO SYSTEM

In the March 1905 issue of *The Pastoralists' Review*, Trew published a description of what he called his 'lasso system'. He began his account with the following statement:

> *Having for some fourteen years past used a lasso system of my own introduction in the branding of cattle and the working of cattle stations with unvarying success, it has occurred to me that, were the method pursued in working it made known, some of my fellow cattlemen would benefit very materially by it, and the long, tedious journeyings to the yards for the purposes of branding, which many of their cattle have to undergo, would be averted.*[1]

He then went on to provide a highly detailed description of his 'lasso system', the circumstances in which it was developed and the results he obtained, supplemented with drawings and photographs. The basic points are summarised here.

Trew said he first tried branding in the open while he was manager of Glen Helen station, Central Australia, 'some fourteen years past', that is, in 1891. At this time the station was on the market and the owners wanted him to muster and brand the cattle. However, the country around the station yards was drought stricken and the cattle had moved to distant areas where storms had produced feed, but there were no yards. It was impossible to bring cattle to the station yards and the owners didn't want to spend money on building yards where the cattle were, so Trew decided to try branding the cattle in the open, even though he'd never seen it done.

His initial method was to muster and hold some cattle as though on a drafting camp. While Trew's men held the cattle together Trew roped an unbranded animal and dragged it towards a tree so that the rope passed alongside the trunk. When the beast came close to the tree trunk men on foot pushed or crowded it sideways so that the rope bent around the trunk, and the animal was semi-immobilised. Small calves were easily scruffed and branded but,

> *towards the end, when the sturdy bunyips (large unbranded male calves) had to be dealt with, matters were a little different; I had to ride around the tree until the bunyip went in an opposite direction and then took a turn with the lasso around the tree. Gus's party, with an additional black boy, would charge the bunyip en*

masse. From a spectator's point of view the procedure that then ensued was indeed funny. First Gus and the boys would be on top and bunyip underneath; suddenly the position would be reversed, and the bunyip fairly jumping and bucking on top of his assailants. I could hear the thud of his hooves as they came in contact with the chaps' shins and other parts of their anatomy. One of the boys mooched away with both hands tightly pressed against his "tummy," and Gus stood rubbing his shins, looking as though he were playing a losing game, and swearing vengeance against the bunyip, but returning to the fray with renewed determination, they overpowered the bunyip and branded him. This sort of performance recurred frequently during the afternoon, and afforded much entertainment for us horsemen. Later on I took a turn at the ground-work and wrestling, but altogether failed to find the point of amusement.[2]

From Glen Helen, Trew went to manage Clifton Hills and Goyder's Lagoon, and it was on these stations that he perfected his method.[3] Judging by his article, in 1905 the essence of his system was as follows. Cattle were mustered and held together in a mob on a patch of clear ground. A man on a 'calf horse' (plate 23) with a greenhide 'head rope' (a lasso, also known as a catching rope and later as a bronco rope) attached to a surcingle placed over the saddle and around the horse's chest rode into the mob, roped an unbranded animal and dragged it to a set of posts specially constructed for the purpose. This set of posts consisted of only two tall 'head posts' about 115 centimetres apart and two outer leg rope posts, all the posts being in line as shown in Trew's original drawing (plate 24). One of the leg rope posts was situated sixty centimetres from one head post and the other ninety centimetres from the other head post. There were no rails at all.

After roping a beast the horseman rode between the two 'head posts', and as the animal approached the posts men on foot moved towards it so that it shifted away from them, keeping the rope hard against one post or the other. When the beast was finally dragged close enough it was forced around until the side of its head was hard against the post. It was then leg-roped, the ropes tied to the leg rope posts, thrown, and thus immobilised. As soon as it was under control the head rope was released and while branding and other operations were carried out the horseman returned to the mob to secure another animal.[4]

Trew advised that sets of these posts should be erected at the principal waterholes and drafting camps upon the run, and mentioned how a tree, or a pair of trees, could be used in lieu of one or both 'head posts' (plate 25). He claimed that his system saved time, money and horse-flesh, and that it led to quiet cattle which were easy to muster, drive, yard and truck, and which never rushed at night. At the end of his article he remarked that,

Several station managers disapproved of my system. They had always used yards for branding, and with the conservatism peculiar to their class, were of opinion that what they had used was the only thing that could be used in the working of cattle. However, time has proved that my system is unquestionably the best we could have adopted for the purpose of cleaning, quietening, and keeping in that state a large herd of cattle.[5]

Trew's statements suggest two things. One is the implication that right up until he began his experiment, outback cattlemen were still taking cattle on 'tedious journeyings to yards for the purposes of branding', and the technique of roping cattle in the open from horseback was not being used elsewhere, or was of extremely limited application and thus unknown to Trew. The other is that by the time he published his article in 1905 his technique had been adopted by others. After Trew's article was published a series of letters concerning the usefulness or otherwise of the system appeared in *The Pastoralists' Review*. These throw additional light on the origin of broncoing, as much by what they don't say as what they do, and are therefore worth summarising here.

First, W. Thorold Grant, who was 'town manager' for the owners of Glen Helen when Trew began developing his system there, wrote in staunch support of Trew and went on to say,

How daring the project was can only be appreciated by those familiar with the old style of station management in the district to which he refers, and the hidebound prejudice of the cattlemen peculiar thereto, who had imbibed, with the assistance of much whisky, all his ideas of managing cattle and stations from the precepts of a drunken ex-employee of the late Mr. J.H. Angus. I well remember the sulphurous criticisms of the old hands, who had sometime previously been relegated to the ranks of the unemployed, as to what the crimson results would be when it came to yarding the expletive blanks. When all this was successfully accomplished without the historical fusilade [sic] of whips and revolvers and the accompaniment of raucous shouting and profanity, the critics were, one and all, to use their own phraseology, blanky well paralysed...the owners of Clifton Hills, which was the next management that Trew received after the break-up of Glen Helen, were as well pleased with his methods as I have been.[6]

Subsequently, a man named James Cummings wrote to severely criticise Trew's method, and one can't help but wonder if he wasn't the 'drunken ex-employee' referred to by Grant. Cummings claimed that on Henbury station a man named Parke had devised an identical system, but after a few years had dropped it as impractical.[7] He claimed he'd tried the system himself and that his results were the exact opposite of those claimed by Trew – it made the cattle wild and extremely difficult to muster, whereas after branding in yards the cattle were quite tractable.[8]

Trew refuted Cummings' criticisms, cast doubt upon either Parke or Cummings ever having had a lasso system, and made the following observation:

Evidently the point that Mr. Cummings wishes to figure out is this. If a calf is branded in a yard at a fortnight or six months old, it becomes an educated beast at once, and remains so until the end. If, on the other hand, the same calf had been branded, say, just outside the yard, it at once becomes an unmanageable rager, and remains as such, which is rather an ingenious deduction.[9]

Cummings responded to this letter and claimed that after Trew's system had been applied at Glen Helen the cattle were sold, but a great many couldn't be mustered, the implication being that they'd been made too wild by Trew's lasso system. He also claimed that Gus Elliott, the stockmen who helped Trew apply his idea on Glen Helen, had since become a station owner himself, but didn't use the method.[10]

Grant and Trew both responded to Cummings, in separate letters. Grant stated that the Glen Helen muster Cummings referred to was in fact a great success. He also claimed to possess a letter in which Elliott professed his belief in the lasso system but 'for reasons wide of the merits of the question he cannot make use of it at Horseshoe Bend [Central Australia]'.[11] Trew had the final word in this debate, basically accusing Cummings of straying 'away from the golden rule that made immortal the name of George Washington.'[12]

If any sort of 'lasso system' was already in existence before Trew's account appeared, developed by Madrill, Parke or someone else, it could've been expected that someone would write to *The Pastoralists' Review* and set the record straight. Apart from Cummings' assertion that Parke had tried and rejected an identical system at Henbury station and that he himself had also tried it, no one did so. Grant, who as the former general manager of Glen Helen and other stations was certainly in a position to know, strongly supported Trew's claim to have developed the technique and cast doubts upon Cummings' veracity. It may be that Parke had tried a 'lasso system' before Trew, but Cummings states that Parke didn't persist with it and it therefore seems unlikely that any experiment Parke made laid the foundation of the method that became broncoing. Cummings' statements also add weight to Trew's claim that yarding cattle for branding was the way things were being done on most places up to that time.

Chapter 6

TREW'S INSPIRATION - COWBOYS, VAQUERO'S, GUACHOS, WILD WEST SHOWS?

From the evidence presented in the previous chapter there can be little doubt that it was the roping experiment of Compie Trew that led to the development of the open bronco system, but an obvious question is, did he invent this method by himself, or was he inspired by something he'd heard about or seen?

I have little doubt that many nineteenth century Australians knew from books and newspaper items how North American cattlemen handled their stock on the open range. For example, while travelling to the Kimberley in 1885, G.H. Lamond came across several head of cattle in unstocked country in the Victoria River district, apparently left behind by a drover heading further west. He had no gun with him, but remembered reading about how the Americans threw cattle from horseback by catching their tails and pulling them over. This method had been used by one man in Victoria in the 1850s (described above), but Lamond said that no one he knew believed it could be done. With nothing to lose, on this occasion he tried it, and it worked.[1]

In 1890 a journalist writing in the *Adelaide Advertiser* mentioned in passing the Irish novelist Thomas Mayne Reid.[2] Reid lived in the United States from 1840 to 1849 and after returning to England he wrote many adventure novels and children's stories based on his experiences in the USA. He became one of the most widely read authors of such literature in Europe and the USA, and at least one of his novels, *The Headless Horseman: A Strange Tale of Texas*,[3] has a description of a lasso being used to catch mustangs. The casual mention of Reid's name in the *Adelaide Observer* presupposes that readers of the article would know who he was, and that his books were available here.

Australian cattlemen also would have heard about American, Mexican and South American roping from Australians who'd been to north or south America, or from demonstrations by Americans, Mexicans and Argentineans who came to Australia. Trew could have become aware of roping from horseback from any one of these sources and, in addition, he may have heard about it from a friend, Harry Chapman, who emigrated to the USA in 1885 and who is discussed below.

Americans in Australia and Australians in America

During the American gold rush in 1849, many Australians went to try their luck on the California diggings. Little is known about the majority of these men; one exception is Nathaniel 'Bluey' Buchanan who returned to Australia and went on to become a cattleman and probably the greatest bushmen, drover and pathfinder Australia has ever seen.[4] The art of roping from horseback was at its peak in California in the nineteenth century[5] and it's highly likely that some of the Australian diggers saw the technique in action, but there's no evidence that Buchanan or any of the others who returned introduced the American roping technique here.

When gold was discovered in Australia in 1851, many American diggers came here to try their luck; one study claims that within a few years there were 10,000 Americans on the Australian goldfields.[6] There certainly were enough Americans at Ballarat in 1854 for them to form their own contingent at the Eureka Stockade – the 'Independent California Rangers'.[7] A number of them were veterans of the Mexican War of 1846-47,[8] and undoubtedly some had been cowboys or seen American cowboys or Mexican vaqueros roping cattle and horses. Some of these Americans remained in Australia, but if any of them knew how to use a lasso and used it here there is no evidence that it caught on among Australian cattlemen.

As well as those who joined the gold rush to California, in the nineteenth century a number of Australian cattlemen went to the USA or Mexico to take up ranching. Some had very interesting experiences and then returned to Australia. One particularly interesting example is a man named Austin Mack who was the owner of Temora station (New South Wales) in the 1870s. In 1879 gold was discovered on his property and 20,000 diggers swarmed onto the place where they 'forcibly cut fences, took forcible charge of all tanks [and] helped themselves to mutton'.[9] Driven to abandon the station, Mack decided to join a syndicate in the purchase of a cattle and sheep ranch near Las Vegas, New Mexico:

> *Mr. Mack was elected as managing director, and took with him sufficient staff to carry on the major operations out there, depending on the local-bred "cow-boys" for other work. However, on arrival he found that things were not altogether as they were painted, although the country was all that could be desired, splendidly grassed, naturally watered and fenced. On the other hand, the sheep were very inferior, scab was rampant, the native population were arrogant to a degree, would allow no yards to be erected, everything down to a fowl had to be lassooed [sic], "labour" put themselves on (or off) Mr. Mack had no say in the matter! In the "fall" if the number of calves reared was not satisfactory, the ranch cowboy's duty and pleasure was to disappear for a time, and return from "somewhere" with the requisite number! On top of other troubles a dispute arose as to the validity of the titles, though duly attested by the Government. The advice given was not to fight, but "distribute a little" amongst the objectors.*[10]

The syndicate wouldn't suffer extortion or engage in bribery, but instead sold out, and Mack returned to Australia.

In the 1880s a number of men from Western Victoria took up ranching in Mexico. Among them was a man named McKellar who was,

> *treacherously done to death on the Nacimiento Ranch by a half-caste Indian, a vaquero named Jesus Gazah. The latter ambushed McKellar from a clump of cactus by the roadside, shooting him dead from the back of his horse, in a most cold-blooded manner, as he was returning from the nearby township of Mariposa... It turned out later that the killer had been hired for 30 dollars Mexican (about £5 in English money) to shoot McKellar because of a dispute that had arisen with a neighbouring landholder, a Spanish Don, concerning the boundaries of their respective lands. Gazah afterwards got drunk in Mariposa and gave the show away.*[11]

McKellar's sons and widow remained in North America but two other Victorians sold their holdings to the sons and retired to their homelands in Victoria.

As well as people who travelled between Australia and North America, there were settlers who came to Australia via North America. For example, when the American Civil War broke out a Scotsman named Fred Gibson joined the Union Army where he attained the command of a company and later was badly wounded. Soon after the end of the war he left America and came to Australia where he followed station life 'and gradually worked his way overland to Central Queensland.'[12] Of these three examples, Austin Mack definitely saw the lasso in use and the Victorians in Mexico almost certainly did so. Undoubtedly there were other nineteenth century Australians who went to America and saw the technique in action, and later returned to Australia.

There were also nineteenth century Americans, Mexicans and Argentineans who came to Australia for reasons other than gold. For example, the famous Cobb & Co. coach line was started in 1853 by a group of Americans and many of the early coach drivers were from America.[13] In a collection of anecdotes on the history of Warwick district (Queensland) there is the story of a buckjump riding contest between an Australian stockman and a Mexican on Canning Downs station, Queensland, in the 1850s.[14] The Mexican stockman undoubtedly would have known about using a lasso and perhaps demonstrated it, but if he did the technique never caught on.

Of particular interest is an account of two Argentinean gauchos who were brought to South Australia by a squatter named McDonald.[15] McDonald was later speared by Aborigines and by 1863 the gauchos had made their way to Oak Park, an isolated station on the Queensland frontier about 170 kilometres west of Townsville. They were hired by the manager, a Mr Mytton, and a few days later they were sent to bring in a few beasts from a nearby cattlecamp. On their return Mytton was surprised, 'to see Dumbana riding up to the place, the beast following with a lariat round his horns, his mate following with his lariat round the hind leg'. As a result 'for eighteen months afterwards he [Mytton] never built a stockyard: the two Guachos [sic] lassoing the calves and throwing them,

Mr. Mytton ear-marked and emasculated the males'. While Mytton clearly saw the value of the gaucho's roping skills, there is no evidence that other stockman in the region learnt to use the lasso from horseback, or that the technique caught on anywhere else in the 1860s.

In the late nineteenth century and early twentieth century a number of American and Mexican roping artistes came to Australia with visiting circuses. 'World's Circus' which visited Sydney in December 1888 featured 'America's Greatest Horse Trainer' and his six 'Bronco Horses'. However, the trainer wasn't billed as a cowboy and there's no mention of the lasso being demonstrated.[16] 'Texas Jack' arrived in Sydney in March 1890.[17] His act included 'Lassoing and Riding Wild Horses Lassoing and Riding Wild Steers bareback'.[18] After performing in Sydney he joined 'Harmston's Great American and Continental Cirque' and toured with them until early 1891.[19] There can be little doubt that numbers of Australian stockmen saw Texas Jack's show, but whatever roping demonstrations were seen by Australian stockmen and cattlemen during or before 1890, the 'hidebound prejudice' and 'conservatism peculiar to their class' remarked upon by Trew[20] may have blinded them to its potential. However, the visit of another group of cowboys with a later circus may have had a different outcome.

At the very end of 1890 two circuses arrived in Australia almost simultaneously. One was the famous Doc Carver's 'Wild America' which boasted 'Fourteen Lodges of Sioux Indians, Cowboys, Mexican vaqueros, Half-Breeds, and Lasso Throwers.'[21] Later Carver announced a 'Kangaroo Lasso Throwing Contest',[22] though whether the kangaroos were to be lassoed or were to do the lassoing is unclear! A press report after the first show commented that, 'a feat which will be much admired is the cleverness shown by many of the cowboys in the use of the lasso' and it went on to describe various lassoing events.[23]

The other circus to arrive was Wirth's, fresh back from America with a new 'Wild West Show' (plate 26). This landed in Adelaide in December and the acts included recreations of classic events from American western history – Indians attacking a stage coach and a log cabin, the burning of the cabin, cowboys coming to the rescue, the Pony Express, an Indian war-dance, hanging a horse thief, and so on.[24] Point six in the program was a, 'Lassoing Exhibition by the Cowboys' which was described as 'a successful system introduced by the cowboys to America showing how to handle cattle without the aid of stockyards or any enclosure' (plate 27). After about a month in Adelaide the show moved on to perform in other towns and cities around the eastern seaboard.[25]

Circuses were an extremely popular form of entertainment in nineteenth century Australia[26] and there can be no doubt that many Australian cattlemen saw Wirth's cowboys demonstrating their roping skills, particularly in Adelaide where the show coincided with the Christmas-New Year period and many station workers would have been in from the outback. When these men returned to their stations they would have told others of what they'd seen, and in addition, the various acts in the show were described in local newspapers, some of which would have circulated outback. A reporter with the *Adelaide Observer* was particularly impressed with the cowboys' roping skills, and was struck with the potential that use of the lasso offered in parts of the outback:

Our back-block boundary riders might show them some points with the stockwhip, but the cowboys can "lay over them with the lasso," to use an expressive Americanism; and here arises the question whether the lariat or lasso could not be used with advantage on some of our cattle stations. It appears after all to be a more humane appliance than the cruel, cutting stockwhip, which in the hands of an expert can take a strip out of the hide of a wild steer, driving him mad with pain and fear. Rounding up refractory cattle with the horse and whip is as arduous a task as can be imagined, but with the lasso we should think any beast could be captured and branded in half the time by two men. The point is that on the American plains, where vast herds of cattle roam free, the stockyard is not such an institution as it is on our stations, and there are other points which make a difference. Nevertheless the lasso would be a useful adjunct, one would think, seeing how effectual it is in the hands of experts like Captain Sutton, Ralph the Mexican, and Felton. These can send the ready loop over head, leg, horn, or any part they please, and it is a fine sight to see their dexterity—hand, eye, and horse working together with beautiful precision. The lasso is a revelation to Australians, and might come to be adopted here. The nearest approach we have to the lariat is the clumsy roping pole.[27]

If the lasso was already being used for branding in the open on Australian cattle stations when the *Adelaide Observer* report appeared, it might be expected that someone would have communicated with the paper and said so. No letter appeared in the *Observer* advising that cattle were already being roped from horseback, and apparently no one walked in off the street to set the record straight either, because when Wirth's Circus and Wild West Show returned to Adelaide in December 1891 similar remarks were made (possibly by the same reporter):

The riding of the cowboys is first-rate, and their handling of the lasso makes one ask why the handy lariat has not been partially adopted in the plains of South Australia. A horse, when he has once been thrown by means of the lariat, appears to recognise a power he cannot cope with, and the process of "bringing him to his bearings" is then comparatively easy.[28]

These reports, published without later contradiction, add weight to the idea that up to the end of 1891 roping cattle in the open wasn't a feature of Australian cattle work, or was of extremely limited application.

An important point to note is that Texas Jack began demonstrating North American roping techniques in March 1890 and Wirths' cowboys began doing so in December, and Compie Trew apparently began his lasso experiments some time in 1891. Is this mere coincidence? Trew said that neither he nor his offsider had ever seen a beast lassoed in the open. This could mean that neither man had seen a lassoing demonstration at all, or that one or the other had seen such a demonstration but only in a circus enclosure. In any case, from his use of the terms 'lasso' and 'lassoed' it's clear that Trew knew about the roping technique used by cowboys and vaqueros. If he hadn't ever seen it demonstrated he may have read about it or heard about it from someone else, and there's also a strong possibility that he'd been told about it by a 'schooldays' chum', Harry Chapman.

Harry Chapman's father bought Mundowdna station, near Marree in South Australia, in 1860,[29] and when Trew left school in 1880 he went to Mundowdna to work with his mate.[30] Harry must have inherited the station because he sold the property in 1885 and immigrated to America where he took up land in Wyoming in 1888, married a Texan woman in 1889, and started a family.[31] He made a short visit to Australia on business matters in 1892 and while here may well have discussed American roping techniques with Australian friends.[32] Harry appears to have stayed in contact with family members in South Australia and in November 1905 he wrote a letter to an uncle in which he contrasted the cattle-handling methods used in the USA with those used in Australia. The letter was published eight months later in *The Pastoralists' Review* and in it Chapman describes how in Wyoming,

> *They 'work' cattle and sheep...on similar lines to those in Australia. The men who work on the cattle ranches are called cowboys. Why should there be no stockyards (called corrals here) in the vicinity of where they are mustering? They rope the calves on the prairie and then brand them. It saves expense in the matter of 'yarding,' and the work is done just as well. We never cripple any more calves than you would in a yard. My two sons, aged 13 and 14 years, are expert ropers, and when on their holidays from school go out, rope their calves by the hind legs, and drag them to the branding fires...*[33]

In July the following year *The Pastoralists' Review* published another of Chapman's letters, this one written to his old mate, Compie Trew. From reading the letter it's clear that he'd read Trew's article on his 'lasso system' and as well as offering a number of comments based on his experience in America he compared aspects of ranching in Wyoming with his direct experience of station life in Australia. Chapman suggested only one change to Trew's system:

> *The only improvement which I would recommend to your own method of roping cattle is a different saddle. The first thing you or any other Australian would say about our saddles is that they are much too heavy. Putting, however, all prejudice aside, the Wyoming cowboy saddle is the best in the world for both man and beast. The saddle in which I ride myself weighs 40 lbs.! I can rope a 1600 lb. bull and hold him... The great feature in these saddles is that they never hurt a horse's back. Among the hundred head of saddle horses which I here employ you would have hard work to find even one that has ever had the least sign of a sore back. But in Australia the reverse is, and ever was, the case. Again, you can ride a horse further in these saddles than in your light ones, the weight being more evenly distributed. To show my opinion of these saddles, if I were returning to Australia and going into the cow business there again, no man should ride for me who did not use these saddles.*[34]

Chapman went on to describe the American roping technique in some detail:

> *If there is a corral stockyard handy, they cut out the cows and calves, take them to the corral, rope the calves by the hind feet and pull them to the fire. Two boys and a man handle them. The calf being dragged by his hind feet is on his side. The calf wrestler never lets him up. You will find this mode a great deal easier than roping out in the open. We always rope in the open when there is no corral, but even then the boys wrestle the calves down without leg ropes. In roping outside a long rope is used, one about 40 ft. in length; in a corral one that is less than 20 ft. answers the purpose. In a corral the rope is tied hard and fast to the horn of the saddle. You have your calf going straight away from you. Drop your loop just in front of the hind leg; a good roper gets both feet every time. We never rope outside when we can get a corral. Where there is no timber we build wire corrals. We often brand with only two men. One ropes by the hind feet and the other holds the calf down by the head. The roper dismounts, and the horse is taught to hold back. That gives this man a chance to brand, &c. Then, should he want to throw a full-grown steer, and there are two men, one ropes by the head or front feet, and the other by the hind feet, and stretches the animal out. At roping competitions the steer is given 100 feet start. The boy runs up, ropes the steer by the neck or horns, throws the slack of the rope over the steer's back, and lets it come under the belly; then turns his horse sharp at an angle, and throws the steer as hard as possible. The horse keeps on pulling, and the boy jumps off and ties the steer's four feet together (hog-tie) a short rope of some 10 ft., carried for the purpose. It often happens that when the boy goes to tie the steer the horse slacks up, and the steer rises. Then the boy has to get back to his horse as quickly as possible. I have a big chestnut sorrel horse which is very expert. He never slacks up, but just keeps pulling; in fact, when the steer is tied he does not care to be led by the bridle to slack up to get the rope off.*

Chapter 7

WHY IS IT NAMED BRONCOING?

An intriguing question is how the Australian technique came to be called 'broncoing'. 'Bronco' is a Spanish/American word originally meaning 'wild or badly broken horse', but by the 1860s in the United States it had come to be applied, at least occasionally, to any horse.[1] They certainly didn't call their branding technique 'broncoing' and this verb doesn't exist in American English in relation to roping or horses.[2] Although the roping of cattle in the open began in Australia in the early 1890s and the term 'bronco' in reference to an unbroken horse was known around this time,[3] the words 'bronco' and 'broncoing' in connection with the technique only appear to have come into use about twenty years later.

When Compie Trew published his article in 1905 he didn't mention the words 'bronco' or 'broncoing', and neither did the correspondents who commented on his article. Similarly, the words are absent in a 1907 article which describes how to make a twisted greenhide rope for catching and branding cattle in the open,[4] and they are absent from a 1909 article describing what we now know as open broncoing.[5] The earliest known references to the words 'bronco' and 'broncoing' being used to refer to Australian cattle branding are in two documents, both dated 1914. One of these notes that the roping and branding of cattle on Innamincka station in 1913 was called 'broncoing',[6] and the other mentions that Barclay (now Barkly) and Lake Nash stations had small wire yards 'to bronco in'.[7] It therefore appears that in 1905-09 the term 'broncoing' as applied to roping cattle from horseback had not yet entered the language of Australian cattlemen or was of very limited currency, but had come into reasonably widespread use by 1913-14. The absence of the term before this time makes it difficult to trace the origin, evolution and spread of the technique, because whatever system was being used was referred to simply as 'branding'.

It's interesting to note that beneath the caption for photographs of broncoing taken on Rosewood and Newry stations and published in 1928, there is the following: 'The word "broncoing," which is not favoured by many Australian cattle men, is practically a new one in this country, though of long standing in America'.[8] The verb 'broncoing' may have been 'practically a new one' in Australia, but historically it didn't exist in America at all and only the noun 'bronco' was used there.[9]

One possible answer to the puzzle takes us back once again to Wirth's Wild West Show of December 1890. One of the Wirth's Circus cowboys was a young American named George Mellor.[10] George was only eighteen years old when he arrived in Australia, but he'd already spent six months performing with Buffalo Bill's Wild West Show, an indication that he possessed extraordinary riding and other stock-handling skills. Using the show name 'Arkansas Kid' he spent two years with Wirths' circus, demonstrating his rough-riding and roping skills around the eastern states (plate 28). At the end of his contract with Wirths he returned to the USA, but was soon back in Australia, running his own Wild West Show under the name 'Bronco George'. There are records of his show performing in Casino, Tenterfield, Rockhampton[11] and Charters Towers in 1895[12] and Croydon in 1896,[13] and it's likely that he performed in other centres in the eastern states.

Eventually Bronco George became so well known that other performers adopted his name.[14] For example, a 'Bronco George' was interviewed in 1909 by *The Showman*, but his life details show he was not George Mellor.[15] A Bronco George show was performed in Adelaide and Broken Hill in 1906,[16] but it's unknown whether the original Bronco George show operated continuously from 1896 to 1906, so this may have been one of his imitators. As late as the 1940s there was an American calling himself 'Bronco George' performing in Australia, but his real name was George Arthur White.[17]

From the statements of those who saw him in action the original Bronco George apparently possessed a phenomenal skill for riding and handling horses and cattle. One contemporary Australian cattleman described him as the best rider he'd ever seen and possibly the best in the world,[18] and Mellor family tradition has it that the great Lance Skuthorpe voiced similar sentiments.[19] Bronco George married an Australian woman and apart from a second, relatively short period back in the USA (1906-1912), he spent the rest of his life here. He grew sugar cane and raised a family near Mackay, and continued to break in horses, do occasional droving trips and to ride in buck-jump shows.[20] His skills were such that he won a bullock ride when aged 70![21] He died at Mackay in 1952.[22]

From the time he began his own show, Bronco George rapidly became known far and wide for his roping, rough riding and horse breaking skills. As a result, it may be that when Australian cattlemen began using lassoes to catch cattle and drag them out of a mob held in the open, some of them remembered his roping demonstrations. They may have said they were going to 'do 'em like Bronco George' or 'Bronco George 'em', eventually shortening this to 'bronco 'em'. This scenario – admittedly only a guess – would explain how the word came to be applied to the roping and branding of cattle in Australia in spite of its original meaning.

Another possibility is that because of the visiting circuses and Wild West Shows, Australians came to associate the term 'bronco horse' with the horses used for roping. It may be that Australian cattlemen remembered the 'bronco horses' used in these roping demonstrations and applied the term to their own horses when they began roping beasts from horseback and dragging them out of a mob. Other than these suggestions, the adoption of the name 'broncoing' is difficult to explain.

If Australian cattlemen were influenced by the roping demonstrations of Wild West Show Americans and Mexicans, it might be asked why they didn't adopt the western saddle. There are probably a number of reasons. When broncoing first began to take off in Australia it's highly unlikely that western saddles were obtainable or at least, not readily availability, and it was easier to adapt existing saddles and harness. However, Australian saddles with a horn were available in Australia by 1919 and possibly a decade earlier (plate 29),[23] and proper western saddles, made by 'an American trained Saddle hand', were available from at least 1922 (plate 30),[24] but neither ever came into general use among Australian cattlemen. Some stockmen considered the horn to be dangerous, believing that if the horse fell and rolled on the rider the horn could cause serious injury. Others believed that a rope attached low down on the side of the horse made it easier for the horse to pull, but the 'conservatism peculiar to their class' that Compie Trew said was a feature of cattlemen in his time was almost certainly an important factor.

Some present-day cattlemen might be, to quote W. Thorold Grant, 'blanky well paralysed' at the idea that any 'expletive blank' American cowboys or Mexican vaqueros could have been even indirectly responsible for the development of what we call broncoing, and that one of them may have unknowingly lent his 'crimson' nickname to the technique. Historically, among Australian cattlemen there has been, and in some cases still is, a common prejudice against America and things American. At least one old cattleman, Peter 'Pituri Pete' Muir, found this anti-American prejudice difficult to understand, and irritating:

> *For some strange reason Australians in the bush seem to take exception to anyone who effects any Americanism into their garb or way of living. I do agree that Australia has a unique lifestyle of its own and indeed the handling of cattle and horses in Australia's back country had developed along completely different lines to the American style, but both cultures have appeal and I simply cannot understand even today the ingrained bias of some outback people towards things American, yet the same prejudiced stockman will be standing there in an American style sombrero, high heeled boots, wearing a pair of three inch necked spurs with rowels as sharp as a razor, whilst at the same time cursing all things American. Beats me, I just cannot understand their attitude.*[25]

Whatever some Australian cattlemen might think, the evidence suggests that visiting Americans or Mexicans inspired or influenced the development of broncoing, and it is undeniable that the Mexican/American term 'bronco' has been borrowed and become part of Australian English, with a completely new meaning. However, if the idea of lassoing cattle in the open was inspired by the visiting American cowboys (who, of course, had earlier borrowed the idea from the Mexicans),[26] the technique was quickly modified by Australian cattlemen and made uniquely Australian.

Chapter 8

BRONCOING WITH PANELS

Once open broncoing became an established technique it soon came to be used in yards as well as in the open. When done in a yard it was known simply as 'broncoing', and when done without a yard it was called 'open bronco'. There is evidence to suggest that when cattlemen first began broncoing in yards they did it in established drafting yards, and may have used an iron hook tied to a gate post[1] or a forked post set alongside a gateway.[2] In any case, broncoing in yards worked in much the same way as open bronco. After roping a beast the catcher rode through the gate and the procedure continued as if a tree was being used. However, a device was soon invented to take the place of iron hooks or forked posts – the 'bronco panel'.

Bronco panels (also known as a bronco 'rail', 'ramp', 'frame', 'bail' or 'fork') varied considerably in size and construction, depending on the state of the herd or the availability of timber. While individual panels often incorporate features which testify to the ingenuity of the cattlemen, they all have several features in common. All consist, in effect, of two sets of strong posts and rails between 120 centimetres and 170 centimetres high and set end to end ten to fifteen centimetres apart. The gap between the two sections of the panel took the place of the hook or forked tree used in open bronco. On the left side of the gap (as approached by the horseman) the post usually was much taller than the one on the right. This was to prevent the rope from flipping over the gap which could waste time and sometimes could be dangerous for the men on the ground.[3] Each section of panelling had at least two rails (more often three and sometimes four), and was two to four metres long. Sometimes the rails on the right hand side were level and there was an additional sloping rail extending from the end of the panel to the ground (plates 31, 32, 33). In other cases the top rail or all the rails of the right hand section of panelling sloped downward towards the outer end and in some instances the top rail was made from a curved tree trunk which reached right down to ground level (plates 34, 35).

When using a panel the catcher roped a beast and rode around the right-hand, sloping end, and as he passed around he swung inward so that the head rope automatically slid up the rail. The roped animals often bucked and ran from side to side, and it was up to the men on foot to guide it so that it moved towards the centre of the panel. If all went well the rope rode up the rail and fell into the gap between the two sections of the panel (eg, plates 31, 32, 36, 37), and the horseman continued to drag the animal forward until its neck was close up to the gap. At this point the horseman stopped his horse but the strain on the rope was maintained. If the catcher had reason to dismount a good bronco horse would stand and maintain the strain until signalled to move back. As the roped animal approached the panel or when it was actually jammed against the gap, its off-side hind leg was caught with a greenhide 'leg rope', and by pulling on the rope and pushing the animal the footmen made it swing around so that its rear end was facing the left hand side of the panel.

When it was in this position and if it was small enough, one man could push his body against the beast and hold onto the top rail to keep the animal in position while the off-side hind leg was roped and both front and back leg ropes were secured to the leg rope posts, or to the rails of the panel. When this was done the catcher moved his horse back to ease the strain on the rope, and as the beast struggled to escape it quickly fell onto its off-side, leaving the preferred near-side exposed for branding. Once the beast fell or was pulled over it was held down and the head rope released. The man who released the rope flung it over the panel so that the noose wouldn't get caught in the gap, and while the animal was operated on the catcher coiled his rope and returned to the herd to rope another.

At times animals were dragged up to panels by means other than horsepower. Plate 38 shows a motor vehicle being used to drag a bull to a panel, and illustrates just how vigorously some animals struggled against the rope. Plate 39, taken at Llanrheidol station near Winton in 1940 or earlier, shows a beast being pulled to the side of a yard by motor vehicle and in the 1930s on Willowra station in Central Australia, a very arid region, a camel was used instead of a bronco horse (plate 40).

On stations where the herd was well controlled and most of the branders were calves, panels didn't need to be particularly strong (eg, plates 33, 34), but if the herd was poorly controlled and there were many full grown cleanskin cows and bulls, the panels needed to be very strong, and sometimes were massive (eg, plates 31, 32, 36, 41, 42, 43). To help make a panel strong Queensland cattleman Norm Forster always tried to get a curved log for the second rail that was long enough to go right across both sections of the panel. As well as making the panel stronger the bend meant that the rail would curve away from the sloping top rail.[4] Norm explained that some bronco panels have the middle rail butting against the opposite post, effectively closing the gap at this point, and that this helps strengthen the panel (plate 43).[5]

On some examples the sloping rail was on the left side of the panel or the top rail on both sections of the panel sloped. A double-sloping panel usually had a tall central post with a gap on each side, though some had only the two central posts, both flush with the top rail.[6] Lester Cain, an experienced bronco man from western Queensland and now the owner of Swanvale station near Jundah, said that double-sided panels were not used north of his station, but they existed on properties further south.[7] A fine example in a large octagonal wire yard can be seen just south of the Nocundra pub in south-west Queensland (plate 44). It has a panel in the centre so no matter which way the wind blew the cattle could be placed downwind and dust they raised would not blow over the men at the panel. Photographs show that double-sided panels were also sometimes used in the western Victoria River district, and in the Kimberley (plate 45).[8] In another variation, some bronco panels had the two sections angled towards each other to form a shallow 'V' so that when the beast was pulled close to the gap it was semi-enclosed, and closer to the rails (eg, plates 32, 33).[9]

Panels with sloping top rails on both sides better suited left-handed catchers because such men sometimes used an 'off-side rig' – that is, they attached the head rope to the off-side – but it's main advantage was that if a roped beast managed to follow the horseman around the end of the panel, the horseman could continue around the other end of the panel and pull the beast up to the 'wrong' side. Otherwise, the beast had to be hunted back around to the correct side before being dragged to the panel.[10]

Traditionally bronco panels were made of wood, but towards the end of the 'bronco era' they were made from large steel pipes and concreted into the ground (plate 46). While panels were the preferred fixtures in yards, they were sometimes constructed in the open at convenient locations around the station.[11] Near the end of each section of panel there were sometimes leg rope posts about forty centimetres high (plates 42, 43, 46, 47). Some cattlemen didn't bother with leg rope posts and instead used the posts or rails of the panel itself, but leg rope posts made fastening quicker.[12] Some panels had ratchets or 'rollers' attached to the second or middle rails in place of the leg rope posts (plates 46, 48). Veteran Northern Territory cattleman Buck Buchester said that, 'if you had them Robinson Rollers you could put...a leg rope on a bull and you could put a piccaninny holdin' the rope once you've got it onto the rollers'.[13] 'Culkah' made a similar assessment. On Victoria River Downs in 1895 he used these 'rollers' – 'Robertson's patent stockyard grip' – and said that, 'With one hand you could pull the heftiest Mick's legs taut, and in an emergency hold the rope-end between your knees while castrating.'[14]

When large animals were dragged up to a bronco panel they sometimes struggled violently (eg, plate 38) and if they did there was a danger that the head rope could come out of the gap. To help prevent this some panels had the sloping rail overhanging the gap (plate 43), and old photographs show that some stockmen placed a rope across the gap after the head rope had gone in, to prevent it coming out (plate 31). Charlie Rayment said that on his panels he had wire loops on each main post so that he could slide a rail through them to keep the rope in the slot.[15]

Sometimes a roped animal would throw itself down before reaching the panel, and refuse to get up again. Various ways were used to get it back on its feet, including backing up the bronco horse to slacken the rope and make the animal think it was unrestrained, and sometimes giving the slack rope a flick,[16] but H.G. Lamond claimed that the best way was to jar the head rope. He said that while the bronco horse was leaning into the pull and the rope taut, he would hit the rope with a light stick. The shock was transferred along the rope to the animal and it would usually get back on its feet immediately.[17]

As a matter of interest, while bronco panels weren't used in Mexico, the USA or Canada, it is worth noting that at the turn of the century in Argentina a single set of posts and rails was recommended as a means to immobilise fully grown cattle for immunisation against 'tick fever'. This panel consisted of two large posts about two metres apart, and with two horizontal rails, the top rail being the height of an adult beasts' head and fixed about fifteen centimetres below the top of the posts. The idea was for the beast to be lassoed, the rope drawn over the top rail and the animal dragged to the panel. The extra height of the end posts above the rail prevented the rope from slipping off either end. When the head of the beast was firmly jammed against the panel it was tied to the rail or to one of the posts by a rope around its horns, and it was then inoculated via a vein in its ear (plate 49).

The Argentinean method of immobilising a beast was published in *The Pastoralists' Review* of June 16th, 1903, but whether such a panel had been in use in Argentina for a long time or to what geographical extent it was used is unknown. It's possible that a similar technique was used in Mexico and that Ambrose Madrill demonstrated its use on Blanchewater for a similar purpose, but in Australia the favoured method of inoculating a beast was to use a special 'seating awl' to draw a thread soaked in serum through its tail.[18]

THE EVOLUTION AND SPREAD OF TREW'S SYSTEM

While it's clear that Trew's lasso system laid the foundation for Australian broncoing, there are obvious differences between the equipment he used and the equipment that came into general use throughout much of Australia's cattle lands. His system didn't use a horse collar or the type of breastplate that went around the horse's chest rather than between its legs, it didn't advocate the use of a forked tree or iron hook, and the set-up of posts he recommended was different from the 'standard' bronco panel described above that eventually was used throughout the outback.

When Trew published his article he was forty years old and had been working on cattle stations for twenty-four years (since 1880).[1] If any of these 'standard' features were in more than very limited use anywhere in the outback, it is virtually certain that he would have known about them. Furthermore, Trew was a progressive cattleman and known to have been an innovator. If he'd been aware of any of these features he would have immediately seen their advantages over his own system and adopted them forthwith, and mentioned them in his *Pastoralists' Review* article, or not written it. The fact that he doesn't mention these innovations and that no one who responded to his article mentioned them suggests that these adaptations to his system either were made after his article appeared 1905, or had only been in existence for a very short time. Clearly, once Trew's system became known and spread from station to station, different men saw ways to improve on the technique, and while most of the improvements probably occurred after publication of Trew's article, the 'who, when and where' of them is largely unknown.

Forked Trees and Posts

The earliest evidence for the use of a forked tree in open bronco work comes from an article about Warenda station, near Winton, published in the *Pastoralists' Review* of July 1909.[2] This article also confirms that in the early 1900s broncoing was a new branding method and that previously branding was always done in a yard:

> *'Warenda has adopted the new mode of branding – "Lassoo branding," which is much quicker than the old style of driving cattle to yards. The cows with their*

calves are first cut out from the main mob, and placed in a suitable position for the lassoers to work. The lassoer pulls the calf to where there is a forked post, into this as he passes he slips the head rope, which prevents "calfy" having too much play. The "lassoo" horses keep a tight strain on the rope until the leg ropes are put on and the calf is down, the other end of the lassoo rope being fixed to the saddle by a strong surcingle. In the case of a big rowdy calf, if the lassoer misses placing the head rope in the fork there is trouble, the branding fire, brands and leg ropes are all scattered in every direction, but in the end horse and "lass." win, and the rowdy one becomes subdued, and has to bear the imprint of the well-known station brand.'

As well as forked posts or forked trees, an article written by Alec Marshall indicates that iron hooks were being used in western Queensland at much the same time.[3] At Innamincka station in 1913 a forked tree was used for open bronco work while an iron hook tied to a gatepost was used in yard work. In the latter instance the catcher roped a beast in the yard, then rode out the gate and the rope was guided into the hook.[4]

Bronco Panels

It's easy to see how Trew's suggested set-up of posts could be modified and adapted to create a modern bronco panel and although nothing certain can be said about who might have done this, a strong candidate is a man named Weston. In the Rockhampton *Morning Bulletin* of April 28th 1906 the following item appeared under the heading, 'A New Method of Branding Calves':

A far western correspondent of the Winton Herald" writes: – "Mr. H.V. Weston, of Glenormiston, is, perhaps, the first man in Queensland or Australia to brand calves by a new method. The cows and calves are cut out in the usual way and put into a yard. Any sort of yard – wire, brush, or stakes – will do for the purpose. Then two cowboys on horseback go quietly amongst them and lasso the calves. The calves are then taken to a trap rigged in the yard and branded in the usual way. The trap is always ready, simple to rig, and everything is easy for men and horses. Besides, the cattle do not get knocked about drafting like they do in the old way. Two hundred and fifty-four calves were branded the first day in six hours and 353 in six hours on the last day. Anyone wishing will be furnished with a plan of the traps and the method of working them by applying to Mr. H.V. Weston, Glenormiston. Over 1500 calves were mustered and branded in this way in two weeks on Glenormiston in March – not too bad after the drought.[5]

A copy of the plan of this 'trap' has not yet been located so it is impossible to be certain that the device was a bronco panel, but there is a strong possibility that it was.

The fact that Barclay Downs and Lake Nash station had small wire yards 'to bronco in' by 1913 suggests that bronco panels may have been in use, but this is by no means certain.[6] Laurie Bain, a senior cattleman who has lived and worked all his life on stations in the Pilbara, Gascoyne and Murchison districts of Western Australia, said there were a couple of bronco panels on Milgun station that were built in 1916. This provides another clue as to when panels first appeared. If panels were in use in this region by 1916, chances are they were in use in Queensland much earlier. The earliest known photograph of a such a panel being used anywhere in Australia was taken in 1923 at Wave Hill station, in the north-west Northern Territory (plate 50),[7] but there is no reason to believe it was invented there. Photos taken in 1923 on Bradshaw station, also in the north-west Northern Territory, show mounted horsemen dragging roped beasts, but no panel is in view.[8]

The Spread of Broncoing

A lot more can be said about the spread of Trew's system and the 'standard' system that it evolved into. Available evidence suggests that it spread rapidly in some areas and slowly in others. Without doubt the conservatism of some cattlemen led them to resist the new method, and it's likely that as station personnel moved from job to job the technique 'jumped' past some stations to places some distance away, with stations in between still using earlier systems. If this was the case the technique effectively spread from more than one location.

Trew began developing his system on Glen Helen station in 1891, and used (and no doubt improved) it on Clifton Hills and Goyder's Lagoon from about 1894 or 1895. During musters near a station's boundaries it was normal practice for stockmen from the neighbouring station to join the mustering team.[9] Stockmen from stations neighbouring Clifton Hills and Goyder's Lagoon would have seen Trew's 'lasso system' in action and no doubt some began using it on the properties where they were employed. Other stockmen who left Trew's employ probably did likewise if the owners or managers of stations where they later worked would allow it.

Evidence that broncoing was unknown in south-west Queensland in 1895 can be found in the station records from Victoria River Downs, in the Northern Territory. In 1895 Bob Watson was manager of Marion Downs, a south-west Queensland station about 420 kilometres north of Clifton Hills,[10] and his brother Bill was then manager of Currawilla, in the same general region.[11] Bob left early in 1896 to take up the management of VRD where he remained until mid-1900,[12] and station records show that under his management yards were essential for branding work to be carried out (see below).[13] In other words, he didn't bring the broncoing technique with him from Marion Downs.

In 1903 Dyson Lacy carried out an inspection of Canobe (Canobie) and Davencourt (Devoncourt) stations, on the Cloncurry River in the Gulf country, and his report makes it clear that on both stations cattle still had to be taken to yards for branding.[14] With respect to Devoncourt station he suggested that a branding yard be built at Bourke Well and said that, 'This is very necessary to save cattle, which, otherwise, would have to travel in 20 miles for branding – doing much damage to cattle and losing calves.'

While broncoing had not been introduced on Canobie and Devoncourt by 1903, it probably arrived there soon after. When former ringer and drover Jack Sammon was working in south-west Queensland in the 1970s a very old local man, Arthur Milson, told him that broncoing was introduced on Marion Downs (Diamantina River) by 'a Yank' in 1901. This claim is interesting in light of C.C. Morton's statement that a Mexican introduced broncoing on Pandi Pandi station in 1900. It should be noted that in the cattle lands of the southern USA there has long been intermarriage between people of Mexican ancestry and people of Anglo ancestry, and many so-called 'Mexicans' were actually from Texas or other border states. As a result it's possible that Australians would call any swarthy-skinned 'yank' a 'Mexican', or perhaps call a light-skinned Mexican a 'yank'.

Jack was also told by another elderly local, Byron Nathan, that broncoing was being used on Springvale station about 450 kilometres north-north-east of Clifton Hills in 1908.[15] Currawilla, about 320 kilometres north-east of Clifton Hills, may have been a bit slower to adopt the technique. Charlie Rayment's father worked on Currawilla for Bill Watson in 1906, and later told his son that at that time branding was still being done in a calf pen.[16] Broncoing was introduced on Nockatunga station, about 380 kilometres east-south-east of Clifton Hills, either by J.M. Hughes between 1904 and 1914, or by his nephew M.L. Hughes some time after he became manager in 1914.[17] Further north, the method was in use at Warenda station near Winton by 1909.[18] It's clear that Alec Marshall's statement that use of the open bronco method was widespread in south-west and western Queensland by 1910[19] is correct, and his claim is supported by Charlie Rayment whose father told him that he was using the bronco method in western Queensland in the same year.[20]

Although broncoing was not being used on Canobie and Devoncourt stations until after 1903, 600 crow-fly kilometres away in Cape York the technique apparently was being used on two stations by about 1904. In an article he wrote in 1953 'Culkah' said that in 1904-05 he was manager of Rokeby and Langhi stations near Coen[21] and that, 'In that style of country, with but two yards that were built for the purpose in the outside country, calves were branded bronco-fashion, out in the open, a poor method with touchy cattle, yet infinitely better than letting 'em go clean-skin.'[22]

Moving further afield, in the Victoria River district and East Kimberley various accounts make it clear that broncoing hadn't appeared there before 1900, and possibly not before 1905. For example, in April 1896 the manager of Ord River station explained that,

> *We must[er] one day and brand the next day. The yards are fitted for drafting cows. The Gins attend to the heating of branding irons, one man lassoos [sic] the calves, and one outside pulls them up to the fence. Two men with [green] hide leg ropes, one on fore and one hind leg, make their ropes fast to pegs, then the rop [sic] rund [sic] the neck is let go and the calves come down with a wallop. The head stockman is then ready with the knife and brands, while another earmarks and puts on a numeral for the year.*[23]

In 1899 on Victoria River Downs one of the yards could hold 2,000 cattle, but there were only three or four yards on the 12,000 square mile (31,000 square kilometre) property.[24] This led the then manager, Bob Watson, to complain to the owners that, 'Insufficient yard accommodation has been a great drawback, necessitating many long drives which wastes much valuable time,' and to request permission to build another branding yard.[25]

Jeannie Gunn, the author of *We of the Never Never,* lived on Elsey station for a year beginning in January 1901 and she had the following to say on the way cattle were then handled:

> *Where runs are huge and fenceless, and freely watered, the years' mustering and branding is no simple task. Our cattle were scattered through a couple of thousand square miles of scrub and open timbered country and therefore each section of the run had to be gone over again and again; each mob travelled to the nearest yard and branded. Every available day of the Dry was needed for the work.'*[26]

In October 1905 an 'old stockman' who had worked on stations between Burketown in the Gulf of Carpentaria, and Timber Creek on the Victoria River, said that the system on all these stations was to put everything through the yards, including the branders.[27] He made no mention of branding in the open, but by the time he wrote his article the technique might have been introduced on some of the stations where he'd previously worked.

VRD was bought by Kidman, Forrest, Emanuel and others in 1900, and Bob Watson was replaced as manager by a 'Kidman man', Jim Ronan.[28] Ronan's long term experience of western and south-western Queensland, western New South Wales and north-east South Australia made him well placed to have seen open bronco work in action if it was being used on many stations in those areas.[29] While there's no evidence that he introduced the technique to VRD, it's quite possible that he did so. Ronan left VRD early in 1904 and his friend Dick Townshend took over.[30] At Ronan's behest, Townshend had come to the district in 1900 to be the Wyndham agent for Kidman, but before this appointment he'd been head stockman on Goyder's Lagoon, so he would've been working for Compie Trew at the very time Trew was developing his 'lasso system'.[31] There can be little doubt that Townshend knew of the technique and he's quite likely to have used it on VRD. A year or so after he became manager a traveller reported that there were 'branding and tailing yards on all the most suitable spots on the run and more are in the course of erection'[32]; quite possibly some of these new yards were bronco yards.

The first reference to open bronco in the Victoria River district suggests that it was being practiced by 1908. In 1910 Drover Burgess, who had been taking cattle from Wave Hill since at least 1908 said that, 'A muster in one place is, of course, out of the question, and the practice is to brand the young stock in lots wherever they happen to be.'[33] By 1916 the open bronco technique appears to have become well established in the Victoria River country. Bill McDonald, who went to work with 'Boomerang' Jack

Brady on Montejinni in about 1916 or 1917, used a forked tree for open bronco work, but with an unusual feature:

> *We were branding cattle which were very wild and hard to muster. We used to mostly open bronco. You muster a lot of cattle together in a mob, the blacks hold them together, you cut a tree down with a fork in it leaving one limb leaning up against the fork. Then one man rides the bronco horse into the mob, ropes his clean skin, drags him up to the fork letting the rope slide up into the fork. Two boys leg rope him, the bronco man releases his strain and goes for another clean skin. The roped beast falls and is branded, ear marked and let loose.*[34]

Further west, another source hints that the method was being used in the west Kimberley in 1905. According to Tom Ronan, when his father arrived in Derby in 1905 to take up the management of Napier Downs there were a lot of Queenslanders among the local cattlemen, and 'the locals were adopting the tried Queensland methods of stock-keeping'.[35] Ronan didn't say what these 'Queensland methods' were, but they quite possibly included the open bronco technique. Ronan also states that when he arrived in Broome as a child in 1910 there was a stockman in the west Kimberley known as 'Broncho Bill'.[36] It would hardly be surprising if broncoing was being used in the Victoria River district or the Kimberley at such an early date because from the beginning of white settlement to the present day, Queenslanders or men trained in Queensland have been a dominant presence in both regions. As a result there have been strong social and economic links, and a constant movement of people and knowledge back and forth across northern Australia which continues to this day.

There is no written information about the introduction of open bronco work in the Pilbara, Gascoyne and Murchison districts of Western Australia, but these areas were not settled by Queenslanders or stocked with Queensland cattle, so links with Queensland were much weaker and the transfer of any cattle technology is likely to have been slower. According to Laurie Bain, the broncoing technique using panels was in widespread use in these areas (north of the Gascoyne River) until the 1920s when most Western Australian stations south of Derby converted to sheep raising. He said that the panels built in about 1916 on Milgun station in the Murchison were constructed by Jack Skeahan.[37] Skeahan was one of the pioneers of the Victoria River-East Kimberley region, arriving there in 1885, and late 1915 or early 1916 he moved to the Murchison district.[38] There can be no doubt that he already knew about bronco panels before he arrived, but whether he introduced them to the Murchison, or indeed, if he introduced the broncoing technique itself, is unknown.

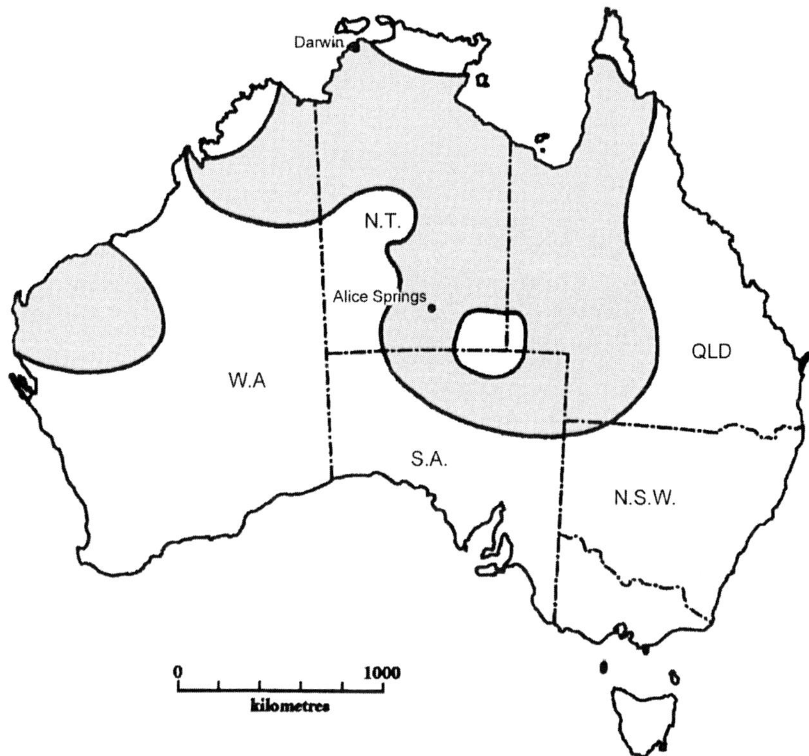

The area of Australia where broncoing was used (approximate).

Alternatives to the 'Standard' Bronco Panel

Writing in 1955, a correspondent to *The Pastoral Review* described a method of broncoing in a post and rail yard that did not require a bronco panel:

> *You can save a bit of space by using two gates in the side of the yard instead of the bronco frame. Each gate needs to be about 5 feet wide and say, a couple of panels apart. The gates should be strongly built, and the second bar from the bottom should extend roughly one foot so as to make a block on the post. The horse brings the calf out of the mob and goes through the gate, which is then closed. The rope is then between the gate and the post, and the calf is pulled up to the bar, which stops it going through. The head rope is taken off and the horse returns to the yard via the other gate.*[39]

Another method was in use on Lakefield station, Cape York Peninsula, in the 1960s. Several bronco yards there had an artificial 'fork' consisting of a stout pole about 130 centimetres long with one end wired to the base of a tree. The pole could be laid flat or lifted upright and fixed to the tree with a wire loop to create a 'dummy post' (plate 51).[40] The pole was laid on the ground until the catcher rode past the tree, and was then lifted and fixed in an upright position to create a 'fork' or gap against which the roped animal was immobilised. The system was simple and obviously of virtually no cost. Sometimes a forked tree within a yard was used instead of a panel (plate 52).

A number of very old yards in the Lake Eyre country have a version of a bronco panel built into the side of the yard itself, instead of a free-standing panel. With this set-up the catcher roped a beast and then rode through an opening (probably a gateway) in the side of the yard. On the left-hand side of this opening (as approached by the horseman) a sloping rail led up to a forked post. After the catcher rode through the gap he would move to the left so that his rope would slide up the sloping rail and fall into the fork. Here it would seem that the idea of using a forked branch on a tree was transferred to the side of a stockyard.[41]

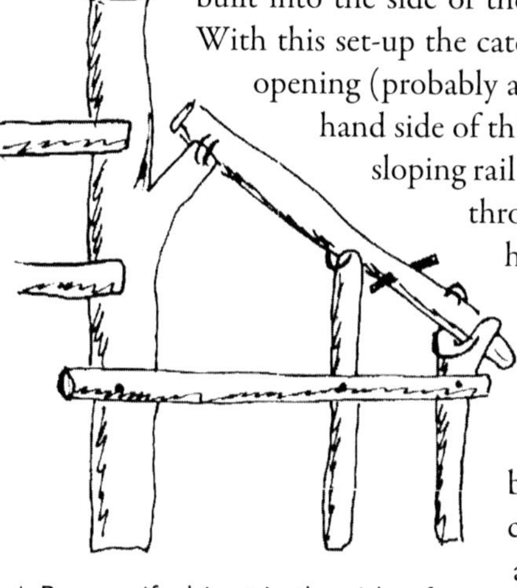

A Bronco 'fork' set in the side of a yard beside a gate (drawing courtesy Philip Gee).

Coincidentally, at the very time that broncoing was taking off in the outback, a new cattle-handling technique was also making its appearance in coastal and near-coastal regions – the calf cradle. This machine was invented by a man named Thomas Bottrell in about 1894 or 1895. According to writer Henry Lamond, there was a small working model of the 'Bottrell crush' at Mt. Cornish station in 1895,[42] and there are photographs and a description of 'Bottrell's Branding Frame' in an article in the *Australasian Pastoralists' Review* of November 1896 (plate 53). The article says that with this frame, 'The branding and castration can then be done with the greatest ease. One man can perform the whole of the work with speed and certainty' and it went on to claim that 'The rough usage attending the ordinary method of performing these operations, which is frequently blamed as a cause of cancer and lumpy jaw in cattle, and consequently loss to the owners, is thus entirely avoided'. Later articles in the *Review* say that Bottrell invented the frame in Stanthorpe, Queensland,[43] and that by 1907 it was already 'largely used on cattle stations in New South Wales and Queensland'.[44] Bottrell's 'Branding Frame' remained in production for decades, but strangely, nearly forty years after it first appeared it remained unknown to some cattlemen.[45]

The branding system used on Gracemere station which used a narrow passage through which calves were forced to be caught by the neck (described above) is clearly a 'race' and bail something like those used on cattle stations in more recent times (see plates 54 and 55). However, there was one difference. Instead of the animals being operated on while caught by the neck and still standing, as is the case with more modern bails, they were leg-roped, their necks released, one side of the crush swung away and they were thrown down for these operations (see plate 9). Bails of the type that held the animal by the neck while all operations were carried out have been in use since at least the 1890s (plate 56).

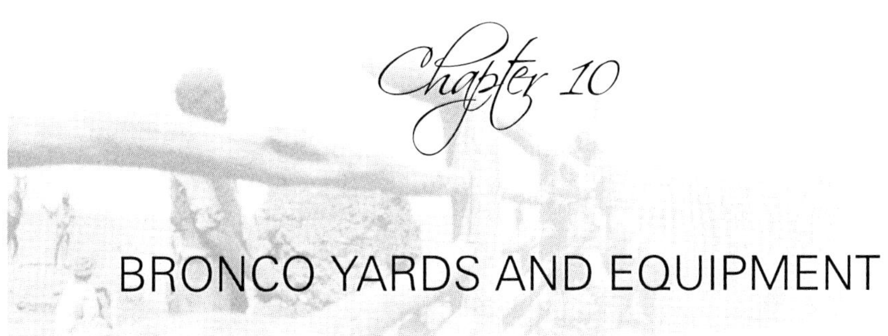

BRONCO YARDS AND EQUIPMENT

The development of broncoing led to various changes in the equipment required to handle the cattle, and in the training of horses and men. As well as the bronco panel discussed above, there were changes in yard design, harness and ropes.

Bronco Yards

When cattlemen began broncoing in yards they quickly realised that a simple yard of wooden posts and heavy wire was sufficient to hold the cattle, and this could be built much more cheaply than a standard drafting and branding yard. In the late 1940s a drafting yard cost £1,200 to £1,500 while a bronco yard cost £120 to £150,[1] and once a bronco yard was built less manpower was needed, so money otherwise spent on wages was saved.

On some south-west Queensland stations hessian or calico was used to make temporary 'portable' yards. According to a letter published in the *Pastoralists' Review* in 1908, at a number of places on Diamantina Lakes station the manager had erected sets of thirteen light posts in a circle, the posts being over two metres tall and seven metres apart. Before mustering began two wires were run from post to post, one strand near the top and one near the ground. Then calico was run from post to post and fixed to the wires with spring hooks. When the branding was finished the wires were taken down, rolled up and left at the yard, and the calico was rolled up and taken for use at the next set of posts.[2] The letter does not mention how the cattle were handled in these yards, but fifty years later Jack Sammon saw the remains of similar yards on the neighbouring Springvale station, and each one had a bronco panel in the centre. Arthur Milson, the same man who told Jack Sammon that broncoing was introduced on Marion Downs by 'a Yank' in 1901, told Jack how the yards described here were used and said they were later converted to wire yards.[3]

In many parts of the north the prevailing winds in the dry season – when most of the mustering and branding was carried out – come from the south-east, so the panels were built on the south-east side of the yard to avoid dust. A few yards had a bronco panel at each end so that whichever way the wind blew the men could avoid the worst of the dust, and large circular yards could be used to similar advantage. Bronco panels also had another use. When a beast was let loose after being branded, earmarked and perhaps castrated, it sometimes tried to attack the men who could find safety by leaping to the other side of the panel (plates 57, 58). In this secondary function it resembled the 'branding panel' that was sometimes used in yards in the middle 1800s.[4] This was not a standard bronco panel but a set of posts and rails against which a roped beast apparently was dragged by manpower for branding and other treatment.

Bronco Harness

According to Henry Lamond, 'There are as many ways of rigging a broncho horse as there are methods of making curry', the common feature being that 'the adherents of each style claim it as the one and only proper way to do it.'[5] However, the basic set-up was to attach a horse collar (plates 18, 46, 59, 60) or a strong leather breastplate to a surcingle which had been placed over the saddle and around the body of the horse (plates 17, 31, 61). Compie Trew's suggested harness used a breastplate that came up between the front legs of the horse. Exactly when the wide breastplate that extends horizontally across the chest of a horse came into use in broncoing is unknown, but according to historian and author Lois Litchfield, horse collars were first used in broncoing by Bill Crombie on Mungerannie station in South Australia.[6]

The head rope was fixed to the surcingle on the near-side of the horse, at about the level of the bottom of the saddle-flap (plates 31, 59, 60). Initially the collar, breastplate and rope were attached directly to the surcingle with buckles, but later Compie Trew modified the surcingle to include a heavy iron ring to which the other items of harness could be attached.[7] Later still some bronco men replaced the iron ring with a rectangular iron 'loop' which made the attachment of the various parts of the harness more efficient.[8] Some bronco men had a swivel permanently attached to the ring on the surcingle while others had the swivel attached to the bronco rope.

Cattleman Lester Cain said that breastplates were often used in packhorse mustering camps because, unlike horse collars, they could be folded or rolled up and carried in pack bags.[9] Another advantage of a breastplate is that it could be adjusted to fit any horse while collars come in different sizes and the one being used may not suit every bronco horse in a camp.[10] However, if the animal being broncoed was large a breastplate could dig into the chest of the horse and cut off its wind (see plate 31). Lester said that on some places a bag with an end cut out was rolled up and used like a collar. This had the same portability as a breastplate but would not cut into the chest of the bronco horse. Charlie Rayment also talked about using a rolled up bag in broncoing:

> *When we all got stuck without bronco gear we'd use a rolled-up bag for a collar. In those days...there was a brand of work-horse mixture, 'Riverina' [brand]...and they were pretty big bags. Bit bigger than the old chaff bags, and stronger, and we'd cut the bottom out of the bag, and roll it up. And then we'd put it over the horse's neck, and it come back like a collar. And then...we just put straps from the bag, back to the rings on the surcingle – quite simple. We tied a bit of hay cord here and*

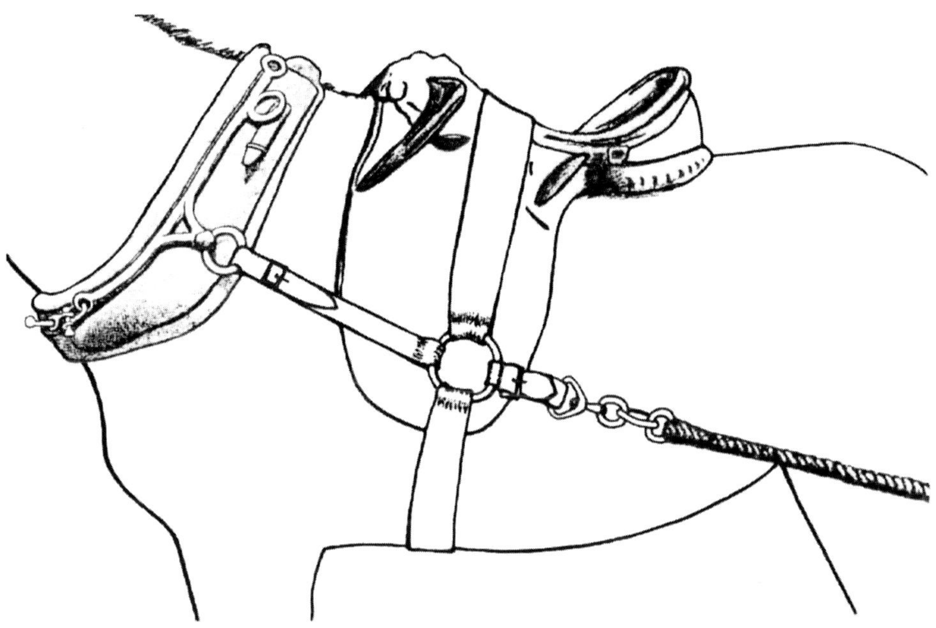

Bronco gear rigged with a collar. In this example the collar and rope are attached to an iron ring but sometimes a rectangle of iron was used, as illustrated below. Depending on the type of hames used on the collar, a chain could be used to connect it to the iron ring instead of a leather strap.

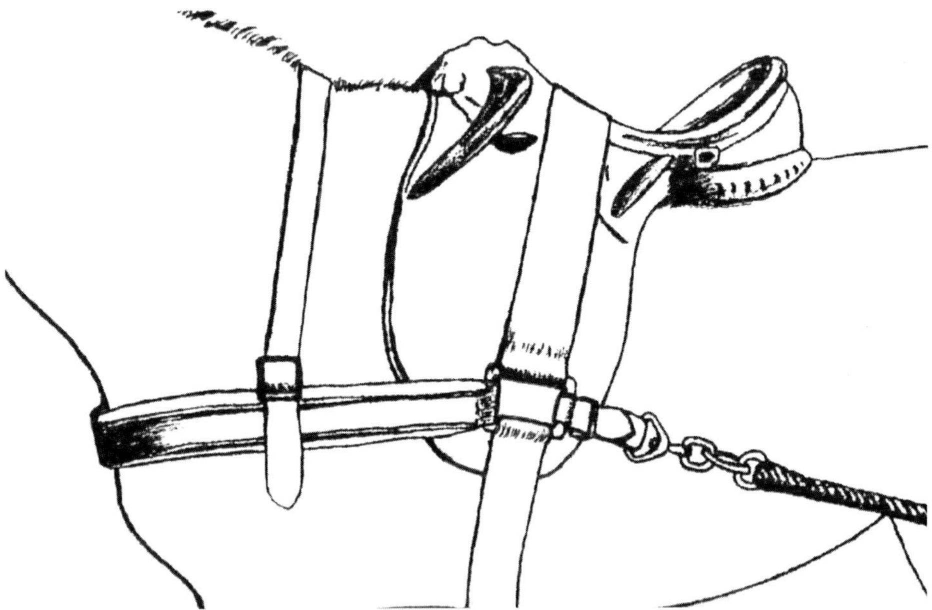

Bronco gear rigged with a breastplate and with a rectangular 'ring' on the surcingle.

> *there on the bag to keep it rolled up tight. ...it's surprising how strong that bag is 'cause, the South Australian fellas, they used it on big cleanskins always, you know. And Jimmy Nunn has told me that down South Australia where he was at Anna Creek they used to get big old canvas mail bags, and they were stronger still, and they'd open the bottom of 'em and roll 'em up and pull 'em over the horse's head, and then tie their side straps onto those bags.*[11]

A point of interest here concerns the way that the Argentinean cowboys or 'gauchos' attached their lasso to their horse. Unlike the Mexican saddle which is derived from the Spanish saddle, the gaucho saddle is a complex indigenous invention consisting of multiple layers of sheep hide, leather and blankets, held together with a broad strap around the horse.[12] It doesn't have a tree, let alone a horn, so the gauchos attached their lasso to an iron ring on the girth strap on the side of the horse, in a manner similar to the way that Australian bronco men attached their ropes (plate 62). However, their rig didn't incorporate a breastplate or collar.[13]

Bronco Ropes

A good strong rope was an essential requirement for broncoing and making them became an important skill and source of pride for cattlemen. Plaited greenhide ropes have been used in stock work for centuries, and certainly were used in Australia since early settlement times. For example, in 1847 Ludwig Leichhardt plaited twenty-five foot (six metre) greenhide ropes to be used for tethering horses.[14] However, with the introduction of broncoing, twisted greenhide ropes largely took over from the traditional plaited version. One reason for this change was that plaited ropes take longer to make, but according to Lamond,

> *unless great care is taken of a plaited rope, and without meticulous care has been exercised in making it, the thing doesn't last any time. It doesn't break or wear out. It loses its life, spring, elasticity, vim, ginger or whatever may be the proper definition. It throws out like a bit of chewed string, and there's no life in it.*[15]

Along with the broncoing technique itself, it appears that Compie Trew also was the first to begin making twisted greenhide ropes to replace the traditional, more time-consuming, plaited version. This was the view of Henry Lamond,[16] of the anonymous writer of Trew's obituary (cited above), and of Trew's contemporary, Artie Rowland, who said that Trew 'was the...first man to start the twisted ropes instead of plaiting them. Twist up the strand first, then put the three of them together. Or you can make it four if you want'.[17]

The earliest Australian description of how to make a twisted greenhide rope for use in broncoing is an article in *The Pastoralists' Review* of August 1907, titled 'Lasso Making'. Writing under the pen name 'O'Grady', the author begins with the following statement:

The making of greenhide lassoes is an art not usually possessed by the Australian cattle man, due no doubt to his method of working cattle in yards where a rough plaited head rope is all sufficient. But for a stockman or cowboy to do good work in the open on a branding camp it is essential that his lasso be light, strong, and pliable.

For the benefit of cattle men who favour branding in the open in preference to employing yards, I will here give an exposition of the procedure pursued in making different sorts of lassoes as dictated to me by an able exponent of the art.[18]

As well as indicating that open branding was still a relatively new thing in Australia at this time, O'Grady's article suggests that using twisted greenhide was a new way of making ropes and provides general support for the proposition that Trew was the first Australian to begin making them. When Charlie Rayment read O'Grady's article he noted the use of the words 'cowboy' and 'lasso', and commented that it was probably written by a 'yank'. While this is certainly a possibility as twisted greenhide ropes were made in Mexico (and South America) and quite probably in the USA, plaited ropes seem to have been more highly favoured among the cowboys and vaqueros.[19] In addition, the terms 'cowboy', 'lasso' and 'ranch' appear in various other writings of the time, and would appear to have been familiar to Australians by the early 1900s.

In the years since O'Grady's article, many descriptions of greenhide rope-making have been published.[20] The basic principals in each description are the same, but there are almost endless variations and differences of opinion as to what makes the best rope. Different ways were developed to prepare and cut the hide, and there were different opinions as to what colour, breed or age of beast was best to use. Different ways were also developed to twist the strands and, subsequently, the rope, and these variations are testimony to the ingenuity and personal experience of Australian cattlemen, as well as to the vital importance of a good rope in broncoing. Indeed, there are far too many variations to be listed here, but the following description combines information from a number of sources to show some of the numerous ways a hide can be prepared and a twisted greenhide rope made.

The first step in making a twisted greenhide rope is, of course, to obtain a hide, and ropes can be made from a hide freshly taken from a beast or from one that had been allowed to dry out. Jack Knox, a ringer in the Victoria River district and East Kimberley in the 1930s, believed that,

Even if the beast has died from natural causes and has lain undiscovered for two or three days, its hide is not only still suitable for twisted rope, but is much more easily flayed. To remove the hide in these cases it is only necessary to make a circular cut from the withers down the shoulder, along the belly and up the flank to the butt of the tail. After turning the beast over, reverse the operation from the butt of tail to the withers. Then hook a horse or vehicle onto the hide near the centre of the belly and the hide can be dragged off free from all foreign matter.[21]

Jack preferred the hide from a white beast, but some believed the hide from a red animal was superior,[22] and yet others believed colour made no difference.[23] Some rope-makers believed a hide from a three or four year old, or a four or five year old beast was the best,[24] but R.M. Williams advocated the hide from an animal less than three years old. Williams also said that a beast that had grown up during drought should be avoided, 'because it will have a papery hide or have bad growth spots on it'.[25]

Some cattlemen believed that the hide from a shorthorn was better than that from a Brahman,[26] and R.M. Williams thought a heifer's hide was better than a steer's and that a bull's hide was 'definitely not good'.[27] Yet another writer said ropes should be made from a hide removed in the winter months.[28] While it may be that one category of hide is better than another, in the opinion of Mick Bower, a former Northern Territory ringer who made plenty of greenhide ropes, there is no guarantee that a particular hide will make a good rope. According to Mick, 'you can get a hide there that makes a beautiful rope, and a hide that looks just exactly the same, and it bloody doesn't!'[29] All rope-makers agree that particular care should be taken not to nick the hide while skinning the beast as this will lead to a weakness in the finished rope, and will shorten its life.

Ropes made from a fresh hide have been described as 'make haste lassoes' because of the short time it takes to make one (two men can make one in two hours).[30] To make such ropes a hide was pegged out as soon as it was taken from the beast and, beginning in the centre, a continuous spiral strip was cut from it (plate 63). This strip was then cut into three or four equal lengths and put through the process of twisting into a rope. A slightly slower version was to salt the hide and fold it up overnight before pegging it out and cutting a strip from it. 'Culkah' described making such a rope by campfire light and using it the next day on 'some fifty-odd wild horses',[31] but others assert that, depending on the time of the year, ropes made from a fresh hide have to be left for a few days or a week to dry out.[32]

More usually ropes were (and are) made from a hide that had been dry-cured, and there are various ways this was done and various claims for the quality achieved. The simplest method was to peg the hide out on the ground, flesh side up, and leave it to dry out. Most writers say this should be done in a shady place,[33] but one man who has made many ropes thought this was 'crap'; he always pegged his hides out in the sun.[34] Some men added salt to the pegged-out hide, but this wasn't necessary, and one cattleman warned that adding salt to the hide during humid weather would cause it to absorb moisture that could make the resulting greenhide too soft.[35] Others advocated sprinkling the hide with salt and folding it up overnight before pegging it out to dry.[36] One writer claimed that the ash of the gidgee tree, found in northern South Australia, central and western Queensland and the south-eastern Northern Territory, was nintey per cent lime and that some cattlemen used it to cure greenhide.[37]

On some stations hides were stretched over and tied to an iron tyre from a large wagon wheel (plate 64). While these were drying crows and ants would clean up any vestiges of meat left on the hide, and when the hides were dry enough a knife was run around the inside of the tyre to produce a circular piece of greenhide. On northern stations these circular pieces were stacked until the wet season,[38] the traditional time for making items such as hobbles and bronco ropes (see plate 65), and a lot of them were needed. For example, Buck Buchester said that each year on the 11,000 square mile (28,500 square kilometre) Wave Hill station he used about twenty hides to make 500 pairs of hobbles[39] (other cattlemen thought twice that many hides would be needed).

'Culkah' believed that winter hides cured the best, but that hides cured in dry, cold, windy weather on the 'north-western plains' (Queensland) were inclined to go 'glassy'. In this area, he said, if hides were cured in the 'early storm season' this effect didn't occur. The view of 'Culkah' was that the hide should be washed immediately after being taken from the beast, then sprinkled with salt, folded or 'bibled' overnight, and pegged out in the morning. He said it should be pegged out in the shade and out of the wind, and after removing any bits of fat or meat, a mixture of fine ashes and salt should be spread over it. 'Culkah' found an old pick-head useful for making the holes on the edge of the hide to take the pegs, and advised that the pegs should be driven in with an outward slant to make the hide as tight as possible. After a week the hide would tighten and then a straight-bladed spade could be used to give the hide a thorough scraping to clean up the inner tissue. At this stage 'Culkah' recommended rubbing a pint (half a litre) of fresh warm milk into the hide because he said it gave the hide a creamy tone and, he believed, helped to soften it.[40]

Before a well-dried hide was cut into strips it had to be moistened until it reached a suitable softness. This could be done by soaking the hide in a creek or a drum. When it was sufficiently soft a strip of the required width, about two to two and a half centimetres, was cut from it. To cut such a strip and maintain a uniform width required a very good eye and a steady hand, or the use of an improvised gauge or a leather plough.[41] If the hide had been dried on a tyre the strip could be cut beginning at the outer edge of the circular piece. Some rope-makers nailed such a hide through the centre to a suitable piece of wood and then turned the hide around as they cut the strip.[42] R.M. Williams described a clever way that a hide nailed through the centre could have the strip to be cut marked on the hide with precision. To do this it was essential that a 'five or six inch nail' (twelve to fifteen centimetres) was used because a nail of this size is the correct diameter for the method to work. Once the hide was nailed down, Williams wound a length of carpenters' or bricklayers' string around the nail, beginning at the bottom and without the coils overlapping each other. Then he tied a black or puce pencil to the string about ten centimetres from the nail and, keeping the string stretched tight and the pencil upright, marked a line onto the hide as the string was unwound. This produced a spiral line on the hide and when the strip was cut following this line it was of uniform width, and the correct width for rope-making.

If a hide was dried while pegged to the ground a strip was usually cut from it *in situ*. To do this, a circle about fifteen centimetres across was first cut from the centre of the hide. A strip was then started from the edge of this hole, with the cutter moving around the hide as the strip progressed. When cutting from a hide pegged to the ground one cattleman said he would start three strips at once, which meant that he didn't have to move around the hide so many times, and he claimed that it shortened the time it took to cut the hide into strips.[43] However, another cattleman said that if you cut three at once you could end up with thick and thin portions of the hide together in the rope, and this would make the rope uneven and possibly weaken it.[44]

When as much as possible of the usable portion of the hide had been cut, one rope-maker said it should be tested by tying it to a post and getting two or three station hands to try and break it by pulling on it – if the strand broke it should be thrown away and another hide used. Some advocated using a sharp knife or a spokeshave to bevel the edges of the strips before they were twisted into a rope,[45] and one suggested passing the strip through a spokeshave to pare back the 'inside shell of the hide'.[46] Most rope-makers didn't bother doing either. Tested or otherwise, the strip was cut into three or four equal lengths, depending on whether a three-strand or four-strand rope was to be made, and then soaked in water until pliable enough to begin twisting into a rope.

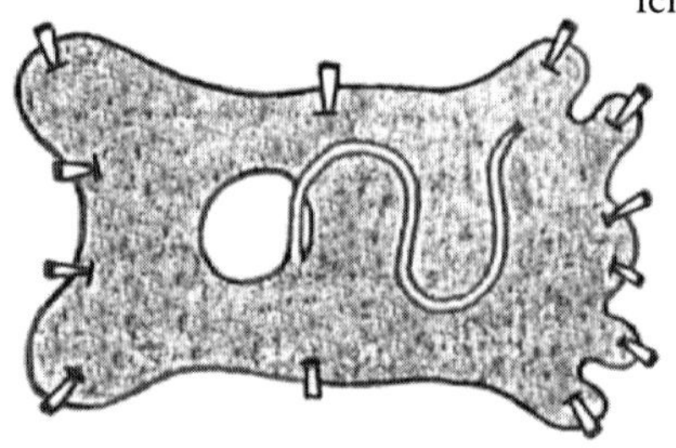

A strand being cut from a pegged out hide, from an article by Jack Knox.

The usual way to make ropes was to first fasten one end of each strand to a single iron ring fixed to a swivel (usually a hobble chain), and to pass the other ends of the strands through a spreader board and tie them to separate swivels. A spreader board is, as the name suggests, a piece of board designed to keep the strands apart and in control while the rope is being formed. It has holes bored through it and for a three-strand rope these holes should form a triangle with sides of eight to twelve centimetres. With a three-strand rope Jack Knox achieved the same effect as a spreader board by placing one strand between the handles of a pair of fencing pliers and the other strands on the outside of the handles. Other rope-makers said that if they were working by themselves they never bothered to use a spreader board or the handles of a pair of pliers, because it was impossible to twist the rope while moving the spreader board or pliers along. Nevertheless, they managed to make good ropes.

The end with the single swivel was then attached to a post or tree and the three swivels at the other end were fixed to heavy weights. Alternatively, the three swivels could be tied to a post or tree about fifteen or twenty centimetres apart, and the other ends of the strands tied directly to a single weight, without a swivel (plate 66). The weight (or weights) used had to be heavy enough to maintain tension on the strands as they were twisted, but light enough to be able to slide along the ground as the tension was increased. 'O'Grady' suggested using flour bags filled with about fifty kilograms of earth,[47] while Charlie Rayment suggested using a coil or two of fencing wire.[48]

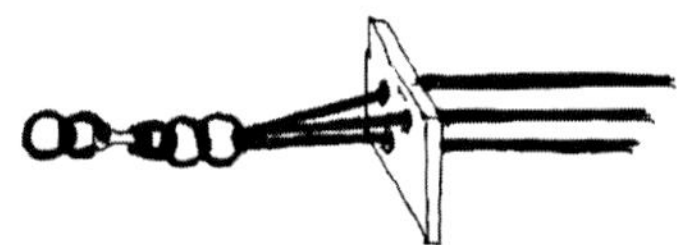

A spreader board with strands attached to a swivel.

To begin forming the rope a long stick or a crowbar was passed through the ring which had the three strands tied to it, and either driven into the ground or laid at an angle so that the ring couldn't turn. At the other end each individual strand was twisted by placing a stick through the ring and turning it around. Each strand had to be twisted more or less the same number of revolutions in the same direction. R.M. Williams said that a strip forty feet (thirteen metres) long should be twisted about two hundred times,[49] while Jack Knox said the strips should be twisted until they showed signs of knotting.[50]

When each strand was twisted enough a longer stick was passed through the ring to prevent the strand from untwisting. When all three strands were twisted they were ready to be twisted together. This could be done by any one of several means. One way was to pass a stick through all three rings to effectively prevent any of them from untwisting and then moving the spreader board to the other end. There, a lever was placed in the ring and the strands twisted together by turning it *in the opposite direction* to that in which the individual strands had been twisted (plate 67). To clarify this point, if the individual strands at one end were turned in a clockwise direction, at the other end they would be seen to be turning in an anti-clockwise direction. Therefore, when strands were twisted together at that end they had to be twisted in a clockwise direction. As the rope formed the spreader board was moved back along the strands, ending up at the end with the three swivels.

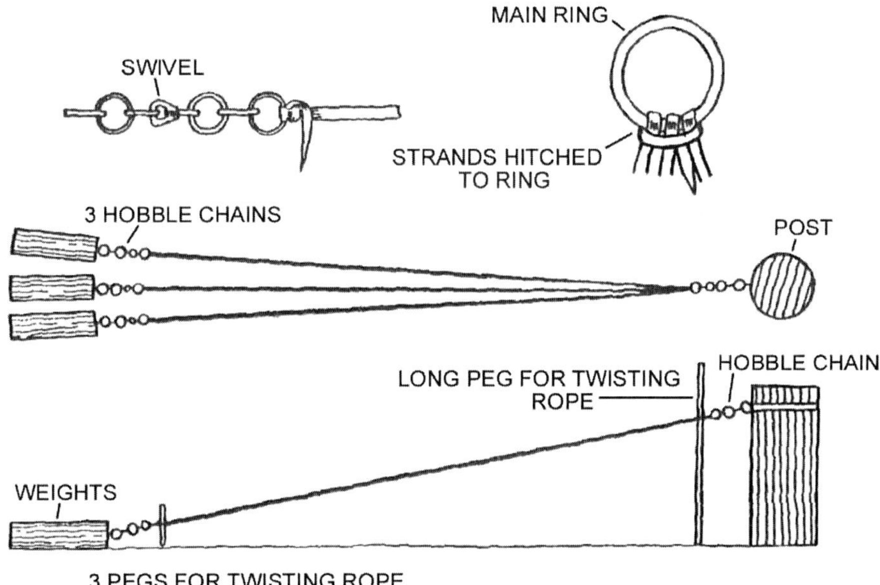

Details of making a rope, adapted from an illustration in an article by Jack Knox. The strands were twisted individually from the left hand (weights) end. The spreader board was then moved close to the post and all three strands were together from the right hand (post) end.

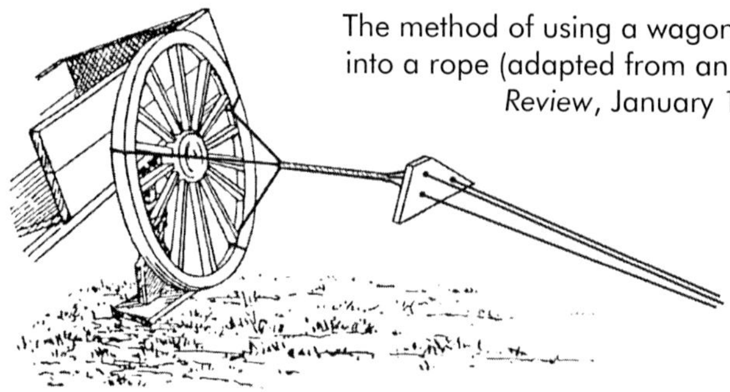

The method of using a wagon wheel to twist the strands into a rope (adapted from an illustration in the *Pastoral Review*, January 16 1923: 46).

If the three swivels had been tied to a post or tree, another way to twist the three strands together was to untie two of them from their swivels and retie them to the third swivel. With the spreader board moved close up this swivel, the ring could then be turned in the opposite direction to that in which the individual strands had been turned, and the spreader board moved back as the rope formed. Yet another way was to untie all three strands and tie them to the outer ends of three spokes of a wagon wheel. The wheel was then jacked up until it could turn freely, and turned so that the strands twisted together.

If the work was to be carried out at a homestead a permanent fixture could be set up for rope making. 'Culkah', a veteran of outback cattle stations from the 1890s onwards when conditions were somewhat primitive, advocated drilling three holes in a large sapling into which were placed green branches shaped like grindstone handles.[51] The ends of the handles were made long enough to protrude beyond the sapling so that the strands could be attached to them, and the handles were turned to twist up the strands, one after the other. A similar but more sophisticated system was to use three metal winding handles set in a short plank fixed to posts set in the ground. The ends of the winding handles either were formed into a hook or had the ends flattened out and a hole bored in them to allow the strands to be attached.

Once the strands were attached to the three crank handles, the loose ends were passed through a spreader board and attached to a single iron ring, swivel and weight, rather than each being attached to its own ring, swivel and weight. The weight could be a coil of fencing wire (about fifty kilograms),[52] a suitably heavy log,[53] a bag of sand or a heavy anvil.[54] Whatever was used for a weight it was crucial that it could slide along the ground as the tension increased. Using the crank handles the strands were twisted individually in the same direction. When all three (or four) were twisted enough the crank handles were tied or in some other way locked together to prevent the strands from unwinding. The strands were then twisted together by placing a stick through the iron ring at the weighted end and using it as a lever. Another method for twisting the strands together was to use a winding handle set in a board, instead of turning the ring with a stick.

In the collection in the Stockman's Hall of Fame in Longreach there is a 'New Era Rope Machine' patented in 1911 and manufactured in Chicago by the Continental Trading

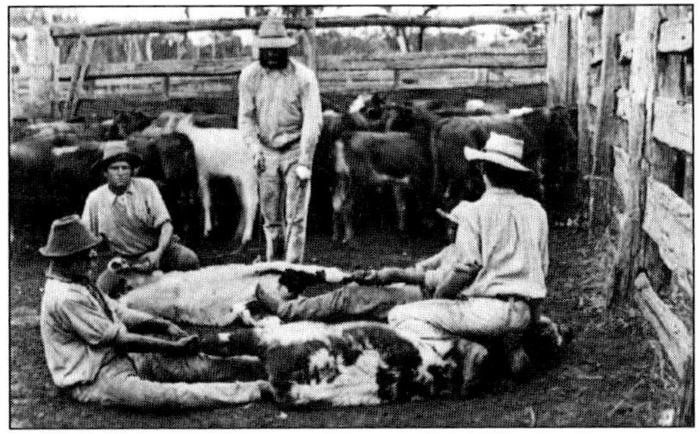

Left

Plate 1: This photograph is available from two sources. A copy in the Oxley Library (Accession no. 91-2-3, neg. no. 14347) has the caption, 'Scruffing calves in Queensland back in the 60s.' A copy published in *The Pastoralists' Review* of February 16 1928 is captioned curiously as, 'The branding: the Samfling method before yards or ropes were used.'

Right

Plate 2: Using the roping pole on a South Australian horse-breeding station

- Illustrated London News

Below

Plate 3: A bull caught with a rope dropped over its horns from a roping pole is being dragged towards a 'sort of a cage' by two men. Another man attempts to force it back with a stick, and meanwhile a beast has escaped by leaping over the rails of the yard.

- Sidney, S. *The Three Colonies of Australia*, Ingram, Cooke and Co., London, 1852

Right

Plate 4. Catching cattle with a roping pole, Canning Downs, Queensland, early 1840s

- Fairholme Collection, Mitchell Library

Left

Plate 5: Dragging a beast to the side of a yard for branding. The rope has been passed under a heavy angled log to prevent the roped beast from moving back and forth across the width of the panel, and a hitch taken around a leg-rope post outside the yard

- Fairholme Collection, Mitchell Library

Right

Plate 6: A painting by S.T. Gill, published in 1865, showing branding in progress in Victoria in the early 1860s. The cattle are being branded from above while the man at the left is forcing unbranded cattle within reach of the branding iron. The cross-bar in the background is part of a gallows used to hoist a carcase up for butchering

- National Library of Australia

Left

Plate 7: Branding cattle from above, Canning Downs, Queensland, early 1840s

- Fairholme Collection, Mitchell Library

Right

Plate 8: For decades Victoria River Downs cattle were essentially wild animals and this photo shows the remarkable ability of some of them to escape from stockyards, a feature noted in the 1840s of cattle in New South Wales. Dashwood yard, c1940

- Walkabout Collection, Mitchell Library

Left

Plate 9: Branding in Queensland in the early 1880s. Cattle were forced down a race at the right and into a gated pen. Once they were immobilised the gates were opened and the animal thrown and operated on

- from H. Finch-Halton's book, *Advance Australia!*, W. H. Allen & Co., London, 1885

Left

Plate 10: Mounted horsemen holding a herd of cattle on a drafting camp on VRD, Northern Territory, a practice that dates back at least to 1860 and probably much earlier

- Gerry Ash Collection

Left

Plate 11: Jim Martin and an Aboriginal stockman pushing a bullock away from the herd after it has been 'cut out', Victoria River Downs, c1940

- Mildred Iverach Collection

Above

Plate 12: Branding at Sturt Creek station, south-west Kimberley, in 1902. The photo shows a horse at the far left being led by an Aboriginal stockman, a calf in the centre of the photo being castrated, and an Aboriginal stockman about to rope another calf at the right. The Aboriginal man at the left is probably holding a leg rope

- R.T. Maurice Collection, Royal Geographical Society of South Australia

Left

Plate 13: Branding on Victoria River Downs in the period 1910-1920. Calves are being roped by hand but pulled to the side of the yard by a mounted horseman. There are leg rope posts outside the rails and a 'dummy' post alongside the big yard post through which the rope is passed to stop it from sliding back and forth along the rail

- John Graham Collection

Right

Plate 14: Open bronco using a forked tree on Alexandria Station, Northern Territory, c1924

- Joyce Galvin Collection

Right

Plate 15: Open bronco using an iron hook tied to a tree, on Keerongooloo station, south of Windorah, Queensland, 1926

- Stockman's Hall of Fame Collection, no. 904

Left

Plate 16: Open bronco on Keerongooloo station, south of Windorah, 1925

- Stockman's Hall of Fame Collection, no. 783

Right

Plate 17: Aboriginal stockman Darby on a bronco horse fitted with a breastplate. He is holding the noose in his right hand while his left hand holds the excess length of the rope and the reins. Victoria River Downs, 1950

- Marie Mahood Collection

Left

Plate 18: An Aboriginal stockman on Argyle station about to give his rope a jerk to tighten it on the roped beast.

- Walkabout photo, 1948.

Right

Plate 19: Mexican vaqueros with a beast roped fore and aft. In Mexico and elsewhere in North America lassos were used, but not bronco panels

- *The Pastoralists' Review*, April 15 1904

Left

Plate 20: Branding calves on a ranch in the west of the USA. The roper waits while a cowboy throws a calf. Two other men hold a calf down while another attends the branding irons

- *Pastoralists' Review*, November 16 1928

Below

Plate 21: Branding cattle on a ranch near Alberta, Canada. This beast has been stretched out after being roped front and back, the traditional way large cattle were handled in North America

- *Pastoralists' Review*, December 15 1905

Below

Plate 22: H. Compton 'Compie' Trew, the originator of what came to be known as 'broncoing'

- *The Pastoralists' Review*, March 16 1905

Left

Plate 23: Compie Trew's 'calf horse'. This is the earliest known photograph showing what is essentially a 'bronco horse'

- *The Pastoralists' Review*, March 15 1905

Below

Plate 24: Compie Trew's recommended system of 'head posts' and leg rope posts, and a pile of wood for the branding iron fire nearby. In this system a horseman dragging a roped beast would ride between the central posts. The idea was that when the animal was almost up to the posts it would wrap itself around one post or the other. In this semi-immobilised state it was then leg-roped and thrown

- *The Pastoralists' Review*, March 15 1905

Left

Plate 25: Compie Trew's men branding a beast. In this instance two trees are being used instead of Trew's recommended head posts. A horseman in the background is probably roping another beast to drag up to the trees

- *Pastoralists' Review*, March 16 1905

Left

Plate 26: Wirth's Wild West Show cowboys, 1890. The figure standing in the white shirt is almost certainly the 'Arkansas Kid' (later to become 'Bronco George'), while the seated man at the right is 'Happy Jack' Sutton, former US Government scout

- Greg Wirth Collection

8 WIRTH BROS.' WILD WEST SHOW

PROGRAMME

OF

WILD WEST SHOW

1. VIRGINIA REEL ON HORSEBACK, by cowboys, Western ranchmen and cowgirls.

2. INDIAN WAR-DANCE, a dance common amongst all aboriginal races. You will note the ferocious expressions on the faces of these North American Indians whilst performing this dance.

3. BUCKING HORSES EXHIBITION by Cowboys and Mexicans.

4. THE EMIGRANT WAGGON, showing the dangers and hardships the hearty pioneer had to contend with whilst using this mode of transit on the desert, long before railways were built, and illustrating an attack by road agents, horse thieves and Indians, who at that time he was always in contest with.

5. THE HANGING OF A HORSE THIEF, an illustration of lynch law in the West, showing how the Westeners saved the United States Government many a dollar in ending a suit in this summary manner. Daring rescue by Minnehaha.

6. LASSOING EXHIBITION BY THE COWBOYS. This is a successful system introduced by the cowboys in America, showing how to handle cattle without the aid of stockyards or any inclosures.

7. EXCITING ATTACK BY THE INDIANS ON THE FAMOUS DEADWOOD COACH, as used in the famous Black Hill gold excitement, 1876. Dashing and daring fight between Indians and passengers. Cowboys to the rescue, thoroughly routing the dusky warriors.

8. COWBOYS' SPORTS. No. 1—Picking up Objects. In illustrating this portion of the programme it is necessary to state that it is portion of the cowboys' education to learn the art of picking up objects, because in lassoing a wild beast they take about three turns of the lasso around the saddle, and if the beast pulls too hard it is compulsory to let the lasso go, and if the boys could not pick up objects they would lose their lasso by the animal's flight, but as it is they do not.

No. 2, Double Picking-up Act for the first time in the world, by Charlie Meadows & Jack Brown. This act has been many times advertised by other shows, but never done.

No. 3, Illustrating how the cowboys protect themselves if their horse should fall, showing how easily they can free themselves from the saddle.

No. 4, COWBOY LIFE.

9. CHASE FOR A BRIDE. Adopted from the Indians, showing the Indians' mode of matrimony in the earlier days in the North-western Indian tribes.

10. THE PONY EXPRESS, illustrating the method of transhipping the United States mail, in early days, across the prairies, before the construction of railways. The rider is supposed to change his horse every ten miles, thus forming a train of horses and the length of his daily journey was one hundred miles.

11. EXCITING CONGRESS OF NATIONS RACE between Mexican, cowboy and Indian.

12. THE ATTACK ON A HUNTER'S CABIN, taken and arranged by Captain Jack Sutton from a scene in real life in the Rocky Mountains.

1. Return of the hunter after a day's shooting.
2. Indians on the trail.
3. The attack by the Indians.
4. Cowboys to the rescue—Indians in difficulty.
5. Firing of the cabin.
6. Exciting Tableau—Death of the hunter—Repulse of the Indians.

WAIT FOR THE CONCERT.

Above

Plate 27: The program for Wirth's Circus, 1890. Item six is a 'Lassoing Exhibition By the Cowboys'

- F.W. Braid Collection

Above

Plate 28: The 'Arkansas Kid' (later 'Bronco George) as he rode off the ship in New Zealand to perform with Wirth's Wild West Show, a few months before he arrived in Australia in December 1890

- Marion Beattie Collection

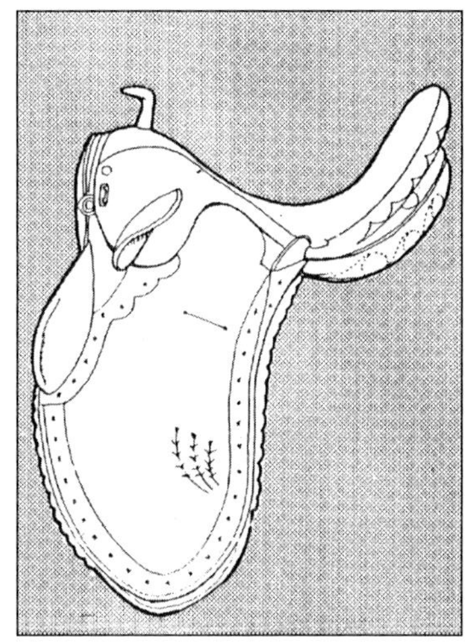

Above

Plate 29: 'Humphrey's Breaking-in Saddle', a combined Australian stock saddle and western saddle

- Courtesy Ron Edwards, *Stockman's Hall of Fame paper*, June 1986, p.12

Below

Plate 30: An advertisement from 1922 showing an Australian stock saddle (left) and an Australian-made American western saddle

- *The Graziers' Review*, October 16 1922

TWO TYPES of

Stock Saddles

THE "WINTON" is the finest Stock Saddle we have yet introduced, and is making a big reputation for us wherever it goes. The flap is sewn direct on to the seat, doing away with the usual Skirt, and giving a narrow non-chafing and most comfortable seat. It is made on our best grade "Fouright" Tree, and the Fittings are Folded Girth, Bevelled Leathers and Solid Nickel Stirrups.

£16/-/-

Write for Price delivered to your nearest Port or Railway Station, as we pay part of the Freight to Queensland, for cash with order.

THE "TEXAS RANGER" is a popular American type Stock Saddle, made by our American trained Saddle hand, equal in every way to the imported article, but much cheaper.

£12/15/-

Write for Price delivered to your nearest Port or Railway Station, as we pay part of the freight to Queensland for Cash with Order.

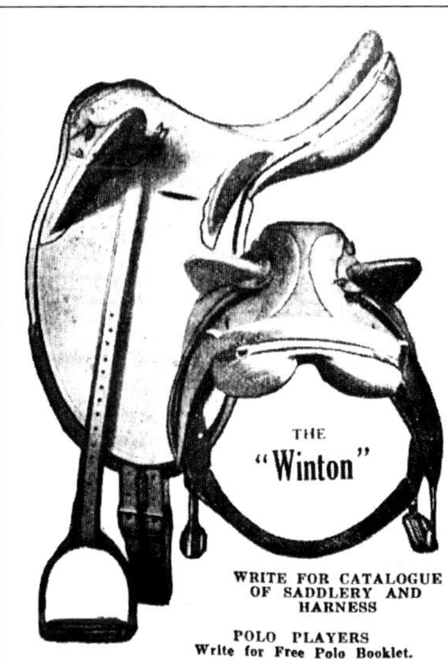

WALTHER & STEVENSON, LTD., Saddlers, 395 George St., SYDNEY

Left

Plate 31: Darby broncoing at Victoria River Downs in 1950. This photo clearly shows how a breastplate can cut into the chest of the horse

- Marie Mahood Collection

Right

Plate 32: A solid angled bronco panel at Chara yard on Nicholson station, East Kimberley, 1957

- Mick Bower Collection

Left

Plate 33: A beast being dragged up to an angled bronco panel on Alroy station on the Barkly Tableland, Northern Territory, in 1966. Timber is scarce on the Barkly so the panel timbers are small, but they are probably made from gidgee which is extremely tough. In any case, the open plains of the Barkly are relatively easy to muster so there would be few large cleanskins to brand and the panel would not need to be especially strong

- John Oxley Library, no. 59701

Right

Plate 34: Dragging a beast towards the bronco panel, Undoolya station, Central Australia, 1961

- F.H. Johnston Collection, National Library of Australia

Right

Plate 35: Laurie Bain broncoing on Woodlands station, Meekatharra, Western Australia, in the late 1960s

- Laurie Bain Collection

Left

Plate 36: Dragging a bull to a panel at Victoria River Downs, 1953. The rope can be seen sliding up the sloping rail and a leg rope man is approaching from the right

- Frank Johnston Collection, National Library of Australia

Left

Plate 37: Dust and smoke rises as a calf is dragged to a bronco panel, VRD, 1953

- Stan May Collection

Right

Plate 38: This spectacular photo shows a bull standing on one leg as it resists being dragged to a bronco panel by motor vehicle. It was taken at Cave Bore on Newry station, Northern Territory, in about 1970

- Bruce Spence Collection, courtesy Bob Young, Brighton Downs, Qld

Left

Plate 39: A 'bronco car' on Llanrheidol station near Winton, Queensland. In this case an iron ring has been set into a rail to act as a guide for the rope, and to stop the roped animal from running back and forth across the width of the rails. There is also a dummy post alongside the post at the left that was probably used for the same purpose

- *The Pastoral Review*, October 16 1940

Right

Plate 40: A 'bronco camel' in a stub yard at Willowra station, Central Australia. Willowra is on Lander Creek which rises in the Tanami Desert

- *The Pastoral Review*, May 16 1942

Left

Plate 41: A bronco panel with massive posts at Nero Yard, Wave Hill station, 1957

- Ellen Kettle Collection, Northern Territory Library

Left

Plate 42: A bull being dragged to a massive bronco panel at Retreat Yard, Victoria River Downs, 1953

- Stan May Collection

Right

Plate 43: Broncoing at Montejinni outstation, Victoria River Downs, c1930. This strongly built panel has the sloping top rail overhanging the gap to help prevent the rope accidentally coming out, and also has the second rail extending across the gap and butting against the other post to give it additional support

- Bobbie Buchanan Collection

Left

Plate 44: A double-sided bronco panel at Nocundra, south-west Queensland 2005. The two sides of the panel have sagged against the central post and closed the gaps where the rope would normally fall into the slot

- Lewis Collection

Right

Plate 45: A beast being thrown in front of a double-sided bronco panel at Argyle station, in the East Kimberley, Western Australia, 1948

- Walkabout photo, Ray Macnamara Collection

Left

Plate 46: Buck Buchester dragging a beast towards a steel bronco panel on Wave Hill station, Northern Territory, 1978. Note the 'Robinson Roller' on the lower rail of the panel

- Mick Bower Collection

Left

Plate 47: A beast leg-roped and with the rope taken around a leg rope post. A man at the left can be seen holding the front leg rope while another man reaches for the tail which he will use to pull the animal over. Victoria River Downs, 1950

- Marie Mahood Collection

Right

Plate 48: A beast being operated on while someone holds a leg rope tight with a 'Robinson Roller'. Barkly Downs, western Queensland, 1960-66

- Lester Caine Collection

Right

Plate 49: A beast tied to a panel after being dragged there by an Argentinian horseman. This was a method used in Argentina to immobilise a beast for inoculation

- *The Pastoralists' Review,* June 16 1903

Left

Plate 50: Broncoing at Nero Yard on Wave Hill station, Northern Territory, August 1923. The catcher is reaching for the rope to coil it after its release from the beast

- AIA Collection, Charles Darwin University, Accession No. 91292. S.C. 1767

Left

Plate 51: A post hinged with wire to the bottom of a tree trunk. The post was laid flat on the ground until the bronco horse dragged a beast past the tree. The post was then stood upright and fixed in place to create an artificial 'fork'. Lakefield station, Cape York, Queensland

- Howard Pearce Collection

Below

Plate 52: John Nicolson and Aboriginal stockmen using a forked tree instead of a bronco panel on Bullo River station, Victoria River district, Northern Territory, in the early 1960s

- John Nicolson Collection

Left

Plate 53: A calf in a 'Botterill Frame', invented by Australian Thomas Bottrell in the early 1890s

- *The Pastoralists' Review*, November 16, 1896

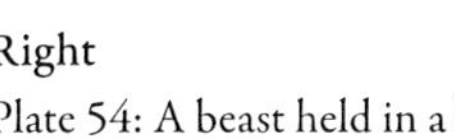

Right

Plate 54: A beast held in a bail while Jim Martin dehorns it with 'parrot-jawed' dehorners, on Victoria River Downs, c1940

- Walkabout Collection, Mitchell Library

Left

Plate 55: A beast being released from a bail on Fitzroy station, Northern Territory, 1997

- Lewis Collection

Right

Plate 56: A beast being dehorned, Benara Estate, Mr. Gambier, South Australia

- *The Australasian Pastoralists' Review*, December 15 1899

Left

Plate 57: Aboriginal stockman 'Daylight' about to leap a double-sided bronco panel to escape a belligerent beast on Argyle station, East Kimberley, in 1948

- Walkabout photo, Ray Macnamara Collection

Left

Plate 58: An Aboriginal stockman leaping the bronco panel to escape a charging, stumbling bull at Montejinni outstation, Northern Territory, c1930

- Bobbie Buchanan Collection

Left

Plate 59: A bronco horse takes the strain with a horse collar, Argyle station, East Kimberley, 1948

- Walkabout photo, Mitchell Library

Right

Plate 60: Brian Crowson with his bronco horse rigged with a collar, Isadore yard, Montejinni station, 1955

- Bert Mettam Collection

Right

Plate 61: An Aboriginal stockman about to throw his rope while men wait at the bronco panel, Victoria River Downs, 1953

- F.H. Johnson Collection, National Library of Australia

Left

Plate 62: The horse of an Argentinian gaucho with lasso attached to a ring on a surcingle on the off-side. This is similar to the way Australian bronco men attach their head rope, but in Australia it is usually rigged on the near-side. The photo was probably taken in the 1930s

- Bovril Collection

Above

Plate 63: Expert bronco man Charlie Rayment cutting a strip from a pegged-out hide, Eildon Park, western Queensland

- Courtesy Charlie Rayment

Above

Plate 64: A hide stretched over an iron tyre to dry in the sun at Mt House station, Central Kimberley, Western Australia, in 1958

- Frank Johnston Collection, National Library of Australia

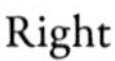

Right

Plate 65: Aboriginal stockmen making greenhide hobbles on Humbert River station, Northern Territory, c1935

- Charlie Schultz Collection

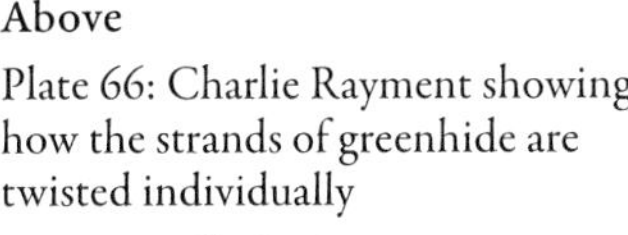

Above

Plate 66: Charlie Rayment showing how the strands of greenhide are twisted individually

- Courtesy Charlie Rayment

Above

Plate 67: Charlie Rayment twisting the three strands of a greenhide rope together

- Courtesy Charlie Rayment

Left

Plate 68: A beast about to be leg-roped as it is dragged close to the bronco panel. The rope being used here has two iron rings – one to act as an eye for the rope and the other to be used to loosen the rope when the beast is being released, Victoria River Downs, 1953

- F.H. Johnson Collection, National Library of Australia

Left

Plate 69: A Mack 711-B road train being tow-started by three bronco horses at Lignum Hole on Camfield station, Northern Territory, in 1967. Only the bronco men have been identified. They are (L-R) Doolya, an Aboriginal stockman from Hooker Creek, Buck Buchester, head stockman on Camfield, and Long Tommy from Pigeon Hole, VRD

- Photo courtesy Mack Trucks Historical Museum, Pennsylvania

Right

Plate 70: At least three bronco men at work in a stub yard on Victoria River Downs, Northern Territory, c1935. Stub yards could be made with lighter and shorter timber than conventional yards, but were more difficult for men to leap over if they needed to escape and angry beast

- Charlie Schultz Collection

Right

Plate 71: Part of an Aboriginal bronco team with a collar for the bronco horse, ropes and a parrot-jawed de-horner. The white man second from the left is a visitor from England. Limbunya station, 1930s

- Sid Bonnell Collection

Left

Plate 72: A bronco panel as sculpture: the Durack Memorial being unveiled at Timber Creek in September 2000. L-R: Tim Baldwin, Mike Reed, Roger Steele and unknown

- Lewis Collection

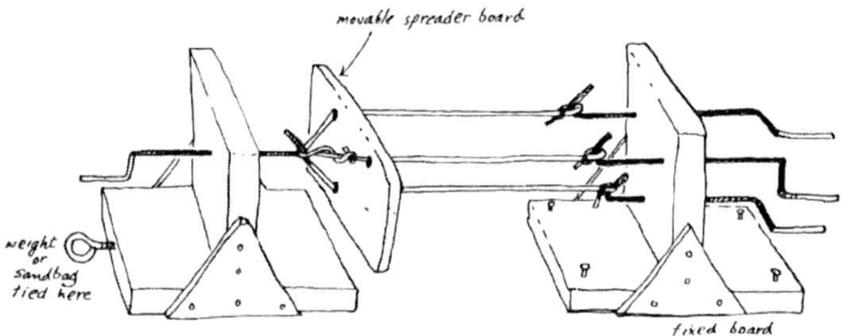

ROPE MAKING MACHINE
A rope making machine, as illustrated in Ron Edwards' book, *Australian Traditional Bush Crafts* (Lansdowne Press, Melbourne, 1975).

A rope-making machine for use at a homestead.

Corporation. Set in a heavy metal housing and with holes at the base to enable it to be bolted to a workbench, this machine has only one handle which turns a series of cogs so that three evenly spaced hooks near the outer edge of the housing and one in the centre revolve simultaneously. To make a rope with this machine, the strands were first tied to the three outer hooks and the loose ends fixed to a swivel and weight, as in the other methods. The handle was then turned so that all three strands were twisted simultaneously. When they were twisted enough the strands were tied to the central hook and the handle was turned to twist them together to form the rope. As with the other methods, the formation of the rope was controlled by a spreader board or the handles of a pair of fencing pliers, as described above. When a rope was finished it had to be left stretched out for a week or more to dry thoroughly. During this time it would sag and this sag had to be taken out by further twisting, or by moving the weight back to increase the tension on the rope.[55]

Once the rope was formed the ends had to be prepared for use. One end needed an eye or keeper of some sort through which the rope could be passed to make a noose, and this was made from greenhide or a heavy iron ring was used. Most rope-makers suggest that the greenhide keeper or iron ring should be fixed to the rope with a 'Turks head' ('wall and crown') knot,[56] but one recommended a 'Mathew Walker' knot.[57] Because the noose could become extremely tight as a roped beast struggled to escape, it was sometimes difficult to loosen. This was especially the case with leg ropes because the noose tightened to a much smaller diameter.[58] It was dangerous to place a hand or finger between the rope and a beast – the beast might move and tighten the rope again, possibly catching the hand and causing a serious injury – so some rope-makers fixed two iron rings to one end of their ropes (plate 68). If this was done, one ring could act as an eye for the rope and the other could be used as a handle to loosen the loop.[59] The other end of the rope was attached either to an iron ring or to a swivel.

It was a matter of personal preference whether the hair was left on the finished rope or was removed. According to 'Q55' in an article he (or she) wrote in 1955, ropes with the hair still on were referred to as 'woolly dogs',[60] but this name does not appear to have ever been in widespread use and may have been restricted to a particular time or place. Most rope-makers recommend removing the hair from a finished rope and there were various ways this could be done. On Argyle station one ringer did it by drawing the rope back and forth through the eye of a mattock head.[61] Another cattleman would get some Aboriginal women to scrape the hair off with butcher's knives, or he'd drag the rope along a dirt road behind a motor vehicle until the hair was worn off.[62]

Some rope-makers suggested greasing a new rope with soap or a thick soap lather.[63] Others recommended using unsalted fat or leather oil,[64] and one advocated a mixture of Stockholm tar and fat.[65] However, 'O'Grady' recommended rubbing a new rope with 'a good sized piece of raw beef (no fat) and plenty of elbow grease' and warned that any other kind of grease would soon destroy it. He said that after such treatment the rope should be briskly rubbed with a piece of chaff bag to remove fragments of meat, and the rope would 'assume a bright lemon colour'.[66]

Greased or otherwise, when bronco ropes were new they were stiff was and it was only after a period of regular use that they reached an ideal suppleness. In between use it was necessary to keep them stretched out, rather than coiled, to prevent a permanent curl developing.[67] Asked how long a rope would last Buck Buchester said, 'Well, about 500 calves I reckon, but [that's] all up – calves and bulls and all.'[68] Given that in their heyday big stations like VRD and Wave Hill branded 20,000 to 30,000 calves,[69] by Buck's estimate they could easily have gone through forty or fifty ropes in a season. Another cattlemen who worked on Wave Hill thought he got much more work out of a rope than this,[70] but the fact is that the life of a bronco rope depended on many factors, including the roughness of the panel being used, the care taken in preventing nicks in the hide while it was taken from the beast, and the size of the cattle being broncoed.

If a bronco rope didn't break first, eventually it became too soft to work efficiently and a common practice was to cut it in half and make leg ropes out of it.[71] In contrast to head ropes which needed to be reasonably stiff or 'springy', leg ropes needed to be quite supple so that they could easily tighten around a beast's leg. Mick Bower used to plait leg ropes, rather than twist them, because 'it's a lot more slippery when you put it on somethin's leg, a beast's leg, well it pulls up tighter and easier.' Mick admitted that it takes more work to make a plaited leg rope, but said that they don't take as much breaking in as one that's been twisted.[72]

Chapter 11

BRONCO MAN AND BRONCO HORSE

Training a horse to do bronco work was another important skill among cattlemen, but apart from a few statements that a bronco horse needed to be strong there is virtually nothing in the literature of a hundred years about what it takes to make a good bronco horse. A strong horse or mule was necessary because on many outback cattle stations there were large or fully-grown cleanskins, and weight and strength was an advantage when dragging such animals to a bronco panel. As a result a heavier type was favoured, 'usually carrying the genes of Clydesdale or Percheron in its breeding.'[1] As Henry Lamond put it, 'A staid stodgy old thing, which doesn't care which way the wind blows, with a fair dash of heavy blood in it, seems to me to be the ideal.'[2]

One cattleman believed that 'any horse will pull', so any horse could be used for bronco work without special training,[3] and this opinion is echoed in Glen McLaren's book *Big Mobs* where one of the sources cited claimed that horses could be taken from the horse breaker to the bronco yard and put to work.[4] However, other bronco men disagreed. Charlie Rayment said that when he was working out near Bedourie, if he wanted to break in a young horse to broncoing he'd start off using a couple of pack surcingles and a few straps. He'd put a double pack surcingle around the chest of the horse and a few straps over the neck to hold it up in place, and he'd catch little calves with this gear.[5] Charlie also said that young horses straight from the breaker needed a lot of training, but he said 'It is possible to get a quiet-natured older horse that has done a few years stock work' to do bronco work.[6]

Lester Cain described the difference between a stock horse and a bronco horse, and also mentioned the need for a bronco horse to be a staid animal:

> *The bronco horses...were different to musterin' horses. Musterin' horses, one person got a horse and...he rode his horse, but the bronco horse, he always had a number of different people ride him, and...that horse had to put up with all the different styles and all the different types, and some of 'em...they got a bit hard in the mouth from the number of different people workin' 'em. Quite often the bronco horses in the camp...became kinda pets. Most people taught 'em to eat bread, and then the bronco horses were kinda considered as pets in the camp, but...I seen some lively times with young bronco horses... When you worked the horses musterin', the horse*

learnt to jump when the cattle jumped, you know, when the cattle moved quick the horse got toey. Well, with the bronco horse you didn't really want that, you wanted a horse that just poked through steady all the time and it was always better if you could get your horse and put him straight into the cattle without givin' him a big lot of stock work. Well that was my opinion of 'em.[7]

Veteran drover Pic Willetts worked on Avon Downs on the Barkly Tableland when the station used a rubber-tyred wagon to break in buggy horses. Later on some of these horses were trained to be bronco horses, but Pic said, 'Buggy horses wasn't always bronco horses and bronco horses wasn't always buggy horses.'[8]

When Buck Buchester was told that some men believed that any horse would pull and could be used for broncoing he strongly disagreed, and he spoke in very forceful and colourful terms about what made a good bronco horse and how he trained them:

Well, I reckon I woulda made more f---n bronco horses than anybody in this country, and nobody can tell me that they can put a green horse in a yard and catch a calf with 'em because they can't. What you used to do, if you got a colt, a green horse, you gonna work him around and pull him around and get him goin' properly before you put him in a [bronco] yard. Now, after pullin' 'im around...I used to put a blackfella up on 'em...and he might ride 'em for a fortnight see, and get 'em goin' properly, just after they come off when I was breakin' in.

Well righto, you take that horse up to the yard. And I used to get on 'im with the gear, and you'd get two blackfellas, and you'd pull 'em around the yard with this horse. Well then, when you brought him back, you'd breechin him - put it around his f---n arse to see if he wouldn't kick or wouldn't f---n lift, and then you'd get on him. And you'd catch three calves and pull the gear off, and then a couple of days after when you were brandin' again you'd bring him back and you'd catch some more calves. You're not catchin' bulls, you're catchin' things about this high – and you gotta learn 'em to pull. A good bronco horse, you can steer 'em around with your knees like this, you're leanin' on this side and they follow you around on the rope.[9]

Buck said he would never let a rider buckle on the bronco rope before he got on a bronco horse, and if a horse got its leg over the rope, he'd train it to lift its leg back over again. Asked how long it would take to fully train a bronco horse, he replied,

To make a top horse out of one? Well, the first year – you don't ride 'em all the f---n year! Like, when you get 'em goin', well you might have 'em there for a fortnight catchin' calves, and then you'd let 'em go – you'd bush 'em then. When you brought 'em back the next year they were a different horse, they knew what you were gonna do...

Three! Three years to make a top horse out of 'em! A lot of 'em, they won't f---n pull at all, they wouldn't pull the hat off your f---n head! ...well I rode a lot of 'em and

a lot of them were no good either. They had no interest in 'em. And another thing, I never ever put a f---n pack on a f---n bronco horse either. Never!'

A well trained bronco horse could be put to uses other than dragging beasts up to a panel or tree. They were commonly used to drag logs to the site where a yard or building was to be erected and on one occasion on Camfield station, three bronco horses were used to start a road train after the vehicle's starter motor failed[10] (see plate 69).

The main skill men had to learn was how to throw the head rope quickly and accurately, and this certainly was not as easy as it looked. Even though the cattle being caught were more or less immobile it usually took some time to become a competent catcher. This was because a bronco camp couldn't afford to let a slow, inexperienced man work for too long as it would hold up the work of branding. Lester Cain described the problem, and how he arranged things in his camp:

When you were young, everyone wanted to get on the bronco horse, but it was fairly hard. There were good catchers there and it takes a long time to learn, and lots of those blokes, they would never, they just had two catchers and put 'em on and they wouldn't let any one learn. But when I was runnin' camps I always give the young blokes a go. I put 'em on, and you never put two on together... When there were plenty of calves when you first started, [you'd] put a learner on with a good catcher and as soon as they get a little bit hard to catch [you'd] put two good catchers up. But I always liked to let everyone have a go to give 'em a chance.[11]

With sufficient practice many men became extremely skilled at throwing the head rope and rarely missed their mark. Lester Cain said that the best catcher he ever saw was,

Jack Elston. He was runnin' the camp on Barkly in the, in the early 60s... Jack would throw a rope and he generally never bothered to even to look back to see whether he caught the calf or not. He just threw the rope and rode to the ramp and never looked back.[12]

Aborigines were often credited with exceptional skill with ropes. In 1926 S.E. Pearson remembered watching a 'half-caste' roping cattle on the Flinders River many years before. He said that, 'amid the pother of dust and angry cattle this yellow-fellow with the 40 ft. of snaking greenhide never missed. Never missed.'[13] Similarly, in the 1940s and 1950s in Western Australia Laurie Bain broncoed cattle on Ethel Creek station at the head of the Fortescue River, and elsewhere in the region (plate 35).[14] In a bronco camp on Ethel Creek Laurie had an Aboriginal stockman doing the catching, and he was exceptionally skilled:

there was an Aboriginal fellow there, absolutely unbelievable! ...we had a yard full of cattle and he went in there this day – Hickey, a boy called Hickey – and he had twenty-one shots and he caught twenty-one calves. And he would catch them and they didn't take more than two steps up to the end of the rope... I mean, some blokes, you know, the calf was there alongside the horse's hind leg. They just dropped the rope on his neck...[15]

Charlie Rayment noted a difference between roping in a yard and roping in open bronco:

> *If I'm catching in a yard, especially a smaller yard, I do not like to...twirl the rope around my head. But if you're open broncoing you've got to catch 'em a little bit further away from you, and you tend to swing the rope so that you can throw it out further. In a yard I...just throw without twirling the rope above my head...but, ah, the good catchers on open broncoing, they all swing the rope, before they throw it.*[16]

The other skill was learning to catch the roped animals with the leg rope. Like head-roping this was just a matter of practice and as Lester Cain commented, men soon learned to be quick:

> *workin' on the leg rope, workin' on the ground is very hard work. If you got two good catchers comin' and those calves are comin' at yer all the time, if you're doin' better than a calf a minute, if you're brandin' close to seventy, a bit over seventy calves an hour...you're certainly goin' on the leg rope. We generally swapped over after about sixty.*[17]

Chapter 12

BRONCO TALLIES

Historical records suggest that the number of animals branded per day has always been a matter of interest and no doubt pride on Australian cattle stations. In bronco camps Pic Willetts worked in the men would,

> *be up there as soon as you could see, lighting the branding fires. Soon as them brands were hot, you got into them. But the men was always racing, to beat one another, to be better than the next fella. The bosses didn't force anybody. They done it out of rivalry.*[1]

It's impossible to compare the different claims made from one branding system to another, or indeed, within one system – this would vary according to the number of men and horses available, the size and wildness of the animals being processed, and probably the degree of truthfulness of those making the claims. Nevertheless, it's interesting to look at some of the claims that have been made.

The branding system used on Gracemere station in the 1870s (described above) could process each calf in less than a minute.[2] On Ord River in 1896 men on foot were roping calves in a yard and men outside were dragging it to the rails where it was thrown and treated. By this method 'Two hundred and eighty to 300 is the record for branding in one day.'[3] Listening to some old hands talk of how they could scruff and treat sixty calves an hour, two young men in Queensland decided to see if they could do better. With the help of men wielding the branding irons and marking knife, they claimed to have scruffed and processed eighty-seven calves in forty-one minutes.[4]

In the early 1900s 'Culkah' was involved in a contest between two stations, with his team using the method where a horse with a rope attached to a swingle bar was led back and forth by a man on foot (described in chapter 2), and the other using a calf-branding crush. He said, 'We used to do up to 80 calves per hour under "competition" conditions, but always over a calf a minute ordinary going, and we kept a bit ahead of the crush record.'[5]

The earliest record of the number of cattle branded per day using the bronco method comes from Compie Trew who said that on the first day he tried his 'lasso system' his team branded only eighty-six,[6] but when he perfected his system he was able to brand 'at the rate of fifty per hour, and for a brush, with handy calves and smart men, sixty'.[7]

Several claims appeared in 1906 in the Rockhampton *Morning Bulletin*. One came from the article about H.V. Weston's 'branding trap' being used on Glenormiston station. Using this device, Weston claimed he was able to brand 254 calves in six hours on the first day and 353 in six hours on the last day.[8] Another reported branding tallies from several different stations, though it is not completely clear that the cattle were being branded by broncoing:

> *Referring to my remarks in a previous issue as to the Brunette records of eighty-six calves branded in sixty minutes, I find that Canobie Station...on the Flinders [River], claims to have put through ninety calves in the same time. Then along comes the Ord Station record of 672 in seven hours, or an average of ninety-six per hour. I take this last record for what it is worth, but judging by the large percentage of "stags" in the Ord River herd, it seems that the male calves were treated as females when this record was being put up. Lissadale [Lissadell] claims to have put through eighty-one in the hour.*[9]

Jim White worked on Brunette Downs in 1920 and claimed that the biggest tally he could recall was 960 in one day. This was achieved by grabbing all the small calves on the edge of the mob in the yard which made it possible to maintain an overall hourly rate of about eighty per hour. This high rate was assisted by the Aboriginal 'boys' on the team who 'were excellent with the ropes and rarely missed a beast as it was broncoed up to the bail.'[10] Len Carrington recalled branding 'over five hundred cattle (I won't say calves) in one day at Fiery Yard using four horses during the day.' He added, 'What a labour saving device the calf cradle is, as some camps made a claim of one hundred and thirty in the hour.'[11] In very large yards there could be three or four horsemen at work at once and consequently a large number of cattle could be treated quite rapidly (plate 70). Alec Marshall wrote in 1937 that in yard work, 'with three ropers in operation...I saw five hundred and forty two calves "broncoed" in a little over four and half hours!'[12]

Most living bronco men can tell of especially high branding rates they achieved. Lester Cain claimed that the most calves he saw branded in a day was on Rokeby station where, 'we done 480 in one day...we had two blokes catchin' at a time',[13] and Central Australian cattleman Ted Fogarty said that, 'If you had a bit of practice, you know...well, you... improve as you go along. On two horses you'd do sixty an hour.'[14] Buck Buchester told me that with three catchers, and by using Robertson rollers and changing his horses at lunch time, 'the most ever I branded, up and down...that's from bloody calves up from bloody three months old up to bulls up to bloody seven or eight years old, the most ever I done was 650.' Buck had heard about a tally on Alroy station of 750, but said, 'You hear of blokes reckon they branded 8 [hundred] or 900 on a bronco panel, I don't believe that one.'[15]

Buck believed broncoing in a yard was the cheapest way of branding a calf and had the advantage that as soon as a calf was let go it went straight to its mother, but he was in no doubt that cradles enabled calves to be branded at a faster rate. With a cradle Buck said that, 'the most ever I branded in a day was a thousand but...I had a bloke draftin' while I was brandin'. Buck went on to say that Peter Sherwin, one of the modern day 'cattle kings', held the record. Sherwin owned properties across the Northern Territory and had brought truck loads of unbranded weaners from a number of his properties to a yard on Walhallow station. There he had six cradles, two head bails and a lot of men. They began branding before sunrise and finished after dark, and in all they branded 10,000 head.[16]

Chapter 13

TALES FROM THE BRONCO YARD

From the foregoing it's clear that broncoing began in Australia in about 1891, and remained a key feature of outback cattle station life for close to a century. In contrast to mustering when stockmen spent much of the day separated and had limited opportunity to talk and interact as a group, broncoing required all the stockmen to work closely together for perhaps days on end. There were undoubtedly millions of individual broncoing events in the bronco era, and while most of the time the work was routine there were occasional humorous events, moments of grave danger or even serious injuries. Most of the old bronco men can tell stories of notable events in the bronco camp or yard. Some of their tales, and some that have appeared in print over the years, are set out below.

Bill Cowan first went outback during World War One. Late in life he wrote about his experiences in western Queensland, including open bronco work on Rocklands station in 1918, at the end of a three year drought. According to Bill,

> *some of the mickies, twelve months and over, used to get a bit savage. After they got to their feet after being processed, they would charge the closest bloke to them and of course there was nowhere to go and you would have to run like hell until one of the ringers on a horse came to the rescue. But we would often see how fast the blokes could run, much to their disgust, before we rescued them. One bloke said to Bill Fernie, the head stockman, 'I don't know what they charge at you like that for when you let them go'. Billie said, 'You don't? Well let me tell you, if you had a greenhide rope around each of your legs and they were pulled apart and you had your knackers cut out and a hot iron on your behind, I reckon you would do the same. Don't you?' 'Cripes', he said, 'I never thought of it that way'.*[1]

On many a bronco camp boys and young men had some fun and perhaps practised for the next buckjump show by riding the smaller beasts as they left the bronco panel. Lester Cain used to do this when he was a boy, but on one occasion he ran into trouble:

> *I got onto a fair lump of a mick and a rode him for about three bucks out from the ramp before I fell off, and then the micky turned round and charged me...everyone else had got on the other side of the ramp, and I raced back and then I couldn't*

jump over the ramp... Brandin' the calves...had dug a bit of a hole in front of the ramp and I fell down into this bit of a hole, and then this mick was there behind tryin' to horn me and I was in the hole and his top horn was bumpin' onto the top [of the bottom] rail. And Jack Alison was runnin' the camp and he was sittin' on the bronco horse waitin' to come up to the ramp, and when they finally got the micky away and I stood up and shook the dust off me, Jack Elston never spoke much, but he said, 'serves you right boy'.[2]

On many stations the cattle were virtually wild animals, and being mustered and driven to a bronco yard did nothing to improve their humour. Of course, the bulls were the most dangerous and most likely to cause problems. Lester Cain had a 'bit of excitement' with a bull at May Downs in western Queensland:

I think it was Billy Abdee on the bronco horse and...he had a bull chase the horse up past the ramp, and he had a bit of trouble. He got around the ramp and then the bull was right onto him and ah, we had a fire made in a forty-four gallon drum, and I stepped in front of the bull to get the bull away from the horse and [then] made a race for this fire, and I jumped over...The bull had hit the end of the rope and broken the rope and he was on one side of the drum and I was on the other side of the drum...and we had a bit of excitement around the drum for a while, till he finally went back into the mob![3]

Alexa Simmons was the 'missus' at Mt. Sanford, an outstation of Victoria River Downs, in the 1950s. On one occasion she took the morning smoko up to the bronco yard, and even though she was heavily pregnant at the time her husband got her to help out in the bronco yard by anchoring the bronco rope in the slot in the panel. She recalled that,

The thing that worried me was, what if a bull jumped the panel before I could get the rope anchored? Can you imagine how I felt? I was just a few feet from an enormous, enraged bull, madly tossing its huge horns and glaring at me between the rails of the panel, its hot breath blowing into my face. Little did the bull know that it only had to surge forward and jump over the rails to be on top of me, but I knew![4]

Sometimes there were serious accidents which couldn't be blamed on the wildness or otherwise of the cattle being broncoed. John Martel was maimed for life through a freak accident in a bronco yard on Strathleven station in Cape York. His wife first heard of the accident on ABC radio and later told the story in her book, *A Christmas Card in April*:

He and two stockmen had been working in a bronco yard when John became entangled in the leg [head] rope which was attached to the bronco horse. Feeling pressure on the rope the horse responded as it was trained to do and pulled John up hard to the bronco post, almost severing his foot. One of the men returned to the homestead to contact the flying doctor as by that time they had a two-way radio at the homestead. However, it was decided that it was too risky to land at the nearby wartime airstrip as it had not been cleared since 1945. Taking the only anaesthetic available – a bottle of rum – the men tied John on to his horse and they rode the three days on to Koolatah where the flying doctor was able to land. By then the leg was gangrenous. He was flown to the base hospital in Cairns and with clever surgery the foot was saved, though at some cost for he was never able to walk properly again and he had to retire to Bettsvale for some years.[5]

As well as danger there occasionally was humour. On Devonport station on the Diamantina River in western Queensland, Charlie Rayment witnessed an event that had elements of both danger and humour, and was amusing to the onlookers if not to the man involved. Lou Bonning (the 'Black Cat') had been breaking in a mare called Little Dumpling and he decided to put her to work in the bronco yard. The saddle he was using was an old military model with the leather cover of the seat missing, and this left two straps of heavy harness leather exposed. These straps were about eight centimetres apart at the front and came together in the middle of the seat in a sharp 'V'. Lou was wearing a pair of blue overalls on which the stitching had come undone in the crotch, and he wasn't wearing any underpants. He was about to get into serious trouble. According to Charlie,

[Lou] gets on Little Dumpling with these overalls on, racin' around the yard swingin' these ropes, and...Little Dumpling dropped her head. She had a few flyin' roots and of course Lou lurched forward and his testicles, they slipped through the wide 'V' in the top of this saddle, and then as he settled back in the seat of the saddle they slipped down where the 'V' come together. He was caught there! Well, the mare kept on rootin' and we could see Lou was changin' colour a bit and lookin' very agitated – we couldn't see what was wrong with him. Anyhow, he stayed there and the mare stopped, the dust settled a bit and he still looked pretty white, so Ray thought he'd better cheer him up a bit. He said, 'Lou, that mare went fairly well, you put up a pretty good ride!' Well, when Lou could get his breath back he said, 'Little mate, if your rocks were caught down in the 'V' in this mongrel bloody saddle you've got me on you'd ride fairly well too![6]

Fortunately for Lou he managed to avoid a personal tragedy. Lester Cain tells the story of another lucky escape on Barkly Downs, on the great plains of western Queensland:

In the 60s...we were...brandin' up at Whistler's Yard. And a bloke was riding a chestnut colt, and he caught the calf on the off side and flipped the rope back up over himself, and when he done that he got a loop caught around his right hand... He stepped off the colt and grabbed it by the cheek strap and hung onto the horse and the calf on the other end was pullin' him out and stretchin' him out a bit, and when we finally got the calf down and got the loop off him he looked at us and said, 'did anyone get a camera and get a photo taken of me while I had my shoulders stretched out wide?' Yeah, he was, he was stretched right out[7]

Many stockmen took pride in their horse skills and their ability to stay in the saddle if a horse bucked, but no one expected a bronco horse to buck. Out mustering on Palparara station in south-west Queensland one day, Charlie Rayment and the head stockman,

Bronty Simms, came across 'Custard', a bronco mare which had escaped the year before and joined a mob of brumbies. They managed to catch Custard and that afternoon Bronty decided to put her to work in the bronco yard. In Charlie's words,

> *Now, he was a hell of a good horseman, Bronty, [but] he did something I would never have done. He caught the mare and put the gear on her – that was the collar bronco gear he had on – and he just got on her and rode her up to the yard at a walk. Didn't give her a lap to warm her up or take any of the kinks out of her. And being below the station, Bronty's wife come down to the yard – Wendy – and we started branding. Well...he caught a calf and Custard dropped her head and rooted very, very well, [and] as she rooted she snapped the reins out of Bronty's hands. Somehow or other something broke on the gear – the collar come over her head or down her neck and there was an awful tangle, and Bronty hit the ground very, very hard, right in front of his wife!*[8]

Charlie reckoned Bronty being thrown from a horse in front of all his men was bad enough, but for it to happen right in front of his wife was doubly embarrassing:

> *Well he never forgave that little mare. He got up off the ground and then he did what he shoulda done first – he pulled the gear off her, jumped on her, took her for a half mile of a gallop down the flat, come back, and put the gear back on. Well that mare worked the rest of the afternoon, I can assure you of that! He pulled everything up! ...Custard certainly earned her bit of grass that night and that afternoon!*

On the vast plains of north-eastern South Australia, trees were often scarce and some became landmarks, the focal point for mustering camps and useful for open bronco work. The single tree on One Tree Plain on Clifton Hills station was such a landmark, and 'Pituri Pete' was working there in the 1940s when this tree met an untimely end. The plan was for the stockmen to muster the plain and bring the cattle to this tree, so the cook had been sent ahead to set up camp and to tie an iron hook onto the tree, ready for broncoing to begin when the cattle arrived. A good mob of cattle were gathered and when they were driven close to where the tree was, no one could see the camp. The men and cattle were in a dense patch of lignum, but normally the tree was visible on the horizon. Eventually they came out onto the grassland, and there was the camp. As Muir described it,

> *I could see a lovely high windbreak made of gum leaves and billies were on the fire, looking the perfect picture of how a stockcamp should look. Away to the right on the far side of the mob which had now come to a halt, was a fresh new looking post, with hook attached ready for the bronco work after dinner...*
>
> *When my turn came to ride in for dinner I learned the truth of the matter. Our smart young ex-serviceman cook had taken his instructions literally as all good soldiers should, he had arrived on camp as scheduled, surveyed the situation and acted accordingly. Taking the axe he first attacked the foliage of the gum tree*

with which he constructed a magnificent windbreak behind the campfire about a hundred yards back from the tree, then after admiring this piece of handiwork he went back to the tree cut the top section of the remaining trunk to post height, cleaned the outside bark from the lower section still standing rooted to the ground, and attached the metal hook thereto. ...he had the satisfaction of knowing he had performed his duty, and he also had the satisfaction of hearing Bill Wilson's tongue at its irascible best.[9]

Chapter 14

FROM BRONCOING TO 'BRONCO BRANDING'

The broncoing technique continued in general use in the outback until the 1970s, but by the end of the 1980s it had been largely replaced by crushes and cradles. It now appears that very few stations still use broncoing for commercial reasons.[1] On Ashburton station in the Pilbara only the homestead yard has a crush so broncoing is still used extensively elsewhere on the property.[2] The technique is still in use on Lambina station in northern South Australia,[3] on Lucy Creek station in Central Australia (seasonal conditions permitting),[4] and on Rosebud station near Mt Isa.[5] Until a few years ago Marion Downs station in the Kimberley combined modern methods with old through the use of a portable yard and a portable bronco panel. The panel was tied to a tree or to a star picket driven into the ground and the portable yard set up around it. If a suitable forked tree was available it was sometimes used in lieu of the panel.[6]

There are a number of reasons why most cattlemen stopped using the technique. One was the introduction of equal wages for Aboriginal stockmen in the mid-1960s (see plate 71). Broncoing is labour intensive and after equal wages came in it became more expensive to use the technique. Lindsay Ward, a Stock Inspector at Halls Creek, reckons that in some areas the introduction of steel yards was another factor because races built of steel could handle big micky bulls that otherwise would kick wooden races to pieces.[7] Probably the crucial factor was the introduction of the Brucellosis and Tuberculosis Eradication Campaign (BTEC Program) in the late 1970s which led to more intensive land use, greatly increased fencing, and most importantly, the need for efficient immobilisation of each animal for disease testing and treatment.[8] On the latter point, in 1985 one authority said that,

> *Though the bronco method has been good enough in a pioneering way for such tasks as branding, castration and dehorning, it has left much to be desired from the point of view of hygiene (dust) and disturbance of cattle during periods of use. Even more important, however, the method has virtually precluded many procedures such as blood sampling, vaccination, treatment of disease or injury, not to mention such advanced techniques as pregnancy testing or weighing, for which a good crush with walk-through bail and drafting yards are minimal requirements if acceptable standards of modern husbandry are to be attained.*[9]

The same author said that the great advantage of open broncoing was that branding could be carried out on the home range of each group of cattle without their having to be moved to a distant fixed yard. This reduced calf losses from failure to mother-up because cows and calves were not separated – a necessity when a calf-branding race and cradle are used. Because of this he believed that the method would probably continue to be favoured for many years! Sadly, he was wrong.

Many old cattlemen lament the change from broncoing in yards and in the open to the use of mechanical crushes. 'Pituri Pete' Muir probably summed up the feelings of many when he commented that,

> *The only drawback that I could see with open bronco was the galloping necessary by the men holding the mob together to keep the cattle on camp. This is exceptionally hard on the horses and they knock up very quickly. Bronco catching in the confines of a stockyard however is very effective and not hard on the cattle or the horses. These days I think the trend is towards ready made steel yards with mechanical crushes, head bails and all sorts of modern innovations which have taken the romantic atmosphere completely out of the cattle game, but I am speaking of the old days when cattlemen were MEN not bloody mechanics, pilots and engineers.'*[10]

The decline of broncoing had become sufficiently obvious by the beginning of the 1980s that a group of Central Australian pastoralists became concerned about the possible loss of the skills involved. To preserve this part of Australia's heritage they decided to try and establish broncoing as a competitive sport.[11] The first 'Bronco Branding' contest was held in Alice Springs in May 1984 and drew competitors from the Northern Territory and South Australia.[12] A demonstration of the new sport was held in Katherine in the Northern Territory in the early 1990s. Quite a number of experienced bronco men were involved, but none of them had used the technique for many years, so their skills were rusty and every one of them missed with their first throw![13]

Interest in the new sport has increased rapidly, with annual contests now held in many centres in Queensland, South Australia and the Northern Territory. The 2004 'bronco branding' National Championship was held at William Creek, northern South Australia, with prize money of about $26,000, and both the number of contests being held and the prize money at the National Championships is expected to grow significantly.[14]

While the skills and techniques of broncoing are now being preserved in the form of a competitive sport, the bronco panel has gone from practical use to museum piece. At the new National Museum in Canberra a display has been set up with a representative bronco panel, a model of a horse wearing a set of bronco harness, and a series of explanatory photographs and texts. The panel has even found a place as a 'sculpture'. In September 2000 a large, steel 'bronco panel' was unveiled at Timber Creek, Northern Territory, as a memorial to the families that pioneered the Northern Territory cattle industry. This is known officially as 'The Durack Memorial', but because it was made by well-known cattleman Ernie Rayner his mates dubbed it 'Ernie's erection' (plate 72).[15]

CONCLUSIONS

Available evidence indicates that for most of the nineteenth century the branding of cattle could only be done in a stockyard, and in many cases this necessitated driving cattle for some days or even a week before a yard could be reached. In earlier decades, once they were yarded cattle to be branded were caught using a roping pole, and dragged by manpower to the side of the yard where they were leg-roped and immobilised. Later a head rope was hand-thrown by men on foot and the roped beast dragged to the side of the yard by manpower, or by a horse being led back and forth by a man on foot.

During the second half of the nineteenth century a number of cowboys, vaqueros and gauchos demonstrated their roping techniques in Australia, and one or two cattlemen may have independently tried a 'lasso system' in the same period. However, there is no evidence that roping cattle from horseback was adopted by Australian cattlemen until the 1890s and early 1900s, and with all due respect to 'gut feelings', until evidence is found to the contrary Compie Trew must be given the credit for developing the roping technique that evolved into broncoing, and came to be used throughout the outback for most of the twentieth century.

And with all due respect to those who are prejudiced against things American and who believe that broncoing is a completely independent Australian invention, the likelihood is that Trew was inspired to carry out his experiments because he knew that cowboys and vaqueros (and perhaps gauchos) used the lasso to catch and brand cattle on the open range. Possibly he was given added impetus to try roping in the open after hearing or reading reports of the amazing roping demonstrations given by Wild West Show cowboys and vaqueros. Similarly, it is also likely the name 'broncoing' that came to be applied to the Australian technique was somehow inspired by visiting Wild West Show cowboys and vaqueros.

It appears that at first Trew's idea was slow to catch on, but by 1905-1910 it was rapidly gaining acceptance, and as it spread various improvements were made. Within about twenty years of his first experiments it's likely that the essential features of broncoing – bronco panels, forked trees or hooks, the use of horse collars or breastplates, and indeed, the name 'broncoing' itself – were established and in widespread use. Trew's technique revolutionised the outback cattle industry, allowing a more efficient and economical use of country with light carrying capacity. It led to new technologies and skills, new terms in the language, and laid the basis, long ago, of the modern competitive sport of 'bronco branding'. While broncoing may be all but extinct as a method for branding on cattle stations today, the annual bronco branding contests will ensure that the many broncoing skills developed in the years since Compie Trew first roped a cow from his horse will be preserved for posterity.

ENDNOTES

Introduction

1 Website: Australian Bureau of Statistics – Australia's beef cattle industry. A million square miles is approximately two and a half million square kilometres.

2 In various historical records the term 'bronco' is also spelled 'broncho'. The word originated in Mexico where it is spelled 'bronco', and in recent times 'bronco' appears to have become the preferred spelling in Australia.

Chapter 1

1 G. McWhiney and F. McDonald, 'Celtic Origins of Southern Herding Practices', *The Journal of Southern History*, vol. 51, no. 2, May 1985: 167.

2 Ibid: 168-69.

3 T. Jordan, *North American Cattle-Ranching Frontiers: Origins, Diffusion, and Differentiation*, University of New Mexico Press, Albuquerque, 1993: 51.

4 Ibid: 51-52.

5 Ibid: 51.

6 Ibid: 52.

7 K.J. Bonser, *The Drovers*, Macmillan and Co. Ltd., London, 1970: 24.

8 Ibid: 25.

9 Ibid: 26.

10 A.R.B. Haldane, *The Drove Roads of Scotland*, Thomas Nelson and Sons Ltd., London, 1952: 8.

11 Ibid: 25.

12 T. Jordan, 1993: 54.

13 Ibid: 51-52.

14 A.R.B. Haldane, 1952: 33.

15 For example, in the early 1960s Rodney Watson had to shoe a number of cows out of a mob that he was walking from Moola Boola station (W.A.) to Queensland (personal communication, Rodney Watson), and Jack Sammon shod two bullocks from a mob he was taking from Alexandria to Marion Downs in 1977 (personal communication, Jack Sammon, June 2005).

16 R.J. Colyer, *The Welsh Cattle Drovers*, University of Wales Press, Cardiff, 1976: 59.

17 A.R.B. Haldane, 1952: 34.

18 Ibid: 50.

Chapter 2

1 J. Thompson and J. Perkins, 'The Wild Cowpastures Revisited', *Journal of the Royal Australian Historical Society*, vol. 77, pt. 4, June 1992: 3-4.

2 B. Fletcher, *Landed Enterprise and Penal Society: A History of Farming and Grazing in New South Wales Before 1821*, Sydney University Press, Sydney, 1976: 36-37.

3 Anon, 'From Most Humble Beginnings: How the First beef cattle came to Australia', *The Australian Shorthorn*, January 1956: 16-18.

4 Anon, *The Romance of the Stockman*, Penguin Books Australia Ltd., Sydney, 1993: 24.

5 J. Carter, *In the Tracks of the Cattle*, Angus and Robertson, Sydney: 71.

6 W.C. Wentworth, *Statistical, Historical, and Political Description of the Colony of New South Wales*, G. and W. Whittaker, London, 1819: 96.

7 P. Cunningham, *Two Years in New South Wales*, Henry Colburn, London, 1827: 290.

8 J. Perkins and J. Thompson, 'The Technology of Open-Range Cattle Farming in Early European Australia', *Prometheus*, vol. 10, no. 1, June 1992: 113-127.

9 G.K.E. Fairholme, Lithograph no. 3, '"Roping" for Branding. Australia', *Fifteen Views of Australia in 1845, by G.K.E.F.*, comprising fifteen hand-coloured lithographs, Mitchell Library, call number ML Q636.2/F; A. Caswell, *Hints from the Journal of an Australian Squatter*, Smith, Elder & Co., London, 1843: 33; A. Harris, *Settlers and Convicts*, Melbourne University Press, Melbourne, 1953: 143-44 (originally published in 1847 by C. Cox, London).

10 J. Hendersen, *Excursions and Adventures in New South Wales,* Saunders and Otley, London, 1854; 'Entertainments. Musical and Dramatic', *Adelaide Observer*, December 27 1890: 26. This deals with Wirth's Wild West Show and the journalist who wrote the article commented that 'the nearest approach to the lariat we have is the clumsy roping pole'.

11 Personal communication, Laurie Lewis, who saw a roping pole being used to catch a horse on Myrtleford station, New South Wales, in the mid-1930s.

12 Letter to the editor, by Arthur Campbell, headed, 'Men and Horses of Today', *The Pastoral Review*, June 16 1951: 624-645. It is unknown how old Campbell was when he wrote this letter, but if he was eighty years old his memories could go back to the mid-1870s.

13 A. Harris, 1953: 143-44.

14 C. Mundy, *Our Antipodes*, Richard Bentley, London, 1852: 289.

15 G.K. E. Fairholme, 1845, lithograph no. 11. 'Branding calves. Australia.'

16 H.W. Haygarth, *Recollections of Bush Life in Australia During a Residence of Eight Years in the Interior*, John Murray, London, 1848: 59.

17 E.O.G. Shann, *Cattle Chosen*, Oxford University Press, London, 1926: 184.

18 M. Cannon, *Life in the Country: Australia in the Victorian Age: 2*, Thomas Nelson, Melbourne, 1973: 116.

19 G.K.E. Fairholme, 1845, lithograph no. 5, 'Australia. Branding cattle in a "pen".'

20 An exceptionally high yard is illustrated in the *Town and Country Journal* of March 6 1875.

21 A. Harris, 1953: 144.

22 C. Fetherstonhaugh, *After many Days*, E.W. Cole, Melbourne, 1917: 60.

23 S. Raymond (ed), *Tourist to the Antipodes: 'William Archer's Australian Journey, 1876-77',* University of Queensland Press, Brisbane, 1977: 37.

24 P.P. Wright. *72 Years in Australia and the South Pacific*, 1919: 3. Mitchell Library manuscript 1452 or microfilm copy CY 1902 (pp. 99-101 of manuscript, frames 066-068 on microfilm). The exact date when this muster occurred is unclear but appears to be between 1858 and 1865. Drafting camps are also mentioned by Oscar de Satge in *Pages from the Journal of a Queensland Squatter* (Hurst and Blackett Ltd., London, 1901: 66), where they were in use on the Darling Downs in 1856.

25 P.P. Wright, 1919: 3.

26 Len Carrington, 'Reminiscences of Augustus Downs During 1920s & 30s', in P. Carrington, *How Many Grids to Gregory?,* Gregory Branch of the CWA, 1977. This method of branding calves was being used on Augustus Downs, in the Gulf of Carpentaria, when Carrington first went there in 1920. According to Jack Sammon, the same method was used on Forest Home station, near Georgetown in the Queensland Gulf Country, as recently as 1965. Jack said that they would draft off the calves, rope them by the back leg and then use a mule to drag them to the side of the yard.

27 H.G. Lamond, 'Branding Methods', *The Pastoral Review*, November 18 1963: 1187.

28 Len Carrington, 1977.

29 'Beginner's Column', *The Australasian Pastoralists' Review*, May 14, 1892: 638-39.

30 *The Macquarie Illustrated World Atlas*, The Macquarie Library Pty. Ltd., Sydney, 1984: 170.

31 Ibid.

32 R. T. Maurice Collection, Royal Geographical Society of South Australia, Mortlock Library, Adelaide.

33 Len Carrington, 1977.

34 D.C. Thompson to the Administrator of the Northern Territory, nd, Australian Archives CRS A3 N.T. 1914/2576.

35 H. Anning, 'Cattle Branding Yards', *The Pastoral Review*, April 16 1931: 358.

36 A. Marshall 'Broncoing!', *Walkabout*, February 1937: 21.

37 H.G. Lamond, *The Pastoral Review*, November 18 1963: 1187.

Chapter 3

1 K. Howard, *Stockman's Hall of Fame paper*, March 2001.

2 Personal communication, Charlie Rayment, Eildon Park station, western Queensland, December 6 2000.

3 Private correspondence, Charlie Rayment, May 20 2002.

4 H. Chapman, 'Ranch Life in Wyoming, U.S.A.', *The Pastoralists' Review*, July 16 1906: 390.

5 J. Mora, *Trail Dust and Saddle Leather*, University of Nebraska Press, Lincoln, 1946 (Bison Book edition, 1973): 55.

6 H.G. Lamond, *The Pastoral Review*, November 18 1963: 1187.

7 A. Delbridge, *et al* (eds), *The Macquarie Dictionary*, second edition, The Macquarie Library Pty. Ltd., Sydney, 1991: 16. The disease is formally known as Actinomycosis.

8 C. Watts, letter to the editor of the *Stockman's Hall of Fame paper*, June 1999.

9 Personal communication, Charlie Schultz, Yankalilla, 1993.

10 Notes taken by Mick Bower of a conversation with Buck Buchester in Katherine, December 16 1999.

11 'J.K. Little' (Obituary), *Pastoral Review*, November 16 1953: 1205.

12 J.K. Little ('Culkah'), 'Our Future Cattlemen', *The Pastoral Review and Graziers' Record*, April 17 1950: 348-49.

13 'Q55', 'Operations at Calf Branding', *The Pastoral Review*, August 16 1955: 1006.

14 Jim White, 'Brunette Downs 1928-45', *Stockman's Hall of Fame paper*, March 2002.

15 H.F. White, 'Branding Performances', *The Pastoral Review*, June 16 1932: 559.

16 F. O'Loughlen. 'Australia Needs More Men Like Jack Kilfoyle'. *The Australian Shorthorn*, August 1952: 31.

17 Private correspondence, Charlie Rayment's comments on the first draft of this book, March 7 2005.

18 Private correspondence, Charlie Rayment, May 20 2002.

19 Personal communication, Charlie Rayment, Eildon Park station, June 12 2000.

20 H.G. Lamond, November 18 1963: 1187.

Chapter 4

1 Geoff Allen, letter to the editor of the *Stockman's Hall of Fame paper*, December 1997: 13. His claim that the term 'bronco branding' was not used until very recent times is supported by the historical record.

2 Personal communication, the late Eddie Hackman, Rockhampton, 1999.

3 *Hoofs and Horns*, November 1957: 57.

4 Personal communication, Charlie Schultz.

5 See also C. Watts, June 1999.

6 E. Morey, 'Timber Creek Patrols, Pt. 5', *Northern Territory Newsletter*, 1978: 16.

7 J.H. Kelly, *Northern Australia Cattle Survey*, Bureau of Agricultural Economics, Department of Commerce and Agriculture, Canberra, 1949 (?): 51-54.

8 Various books, reports and papers written by Helen can be found on the internet.

9 H. Tolcher, *Innamincka: The Town with Two Lives*, Innamincka Progress Association, 1990: 57-58.

10 Personal communication. The original source for this information is the *Adelaide Chronicle*, August 31 1939: 66, column 5.

11 'The Camels and Donkeys', *South Australian Register*, January 27 1866.

12 T. Jordan, 1993: 25.

13 T. Ronan, *Packhorse and Pearling Boat*, Cassell Australia Ltd, Melbourne 1964.

14 T. Ronan, *Deep of the Sky*, Cassell Australia Ltd, Melbourne, 1963 (Australian edition).

15 T. Ronan, 1964: 140-42.

16 N. Pearce, *Homesteads of the Stony Desert*, Rigby, Adelaide, 1978: 29.

17 Timber Creek Police Journal, entries for April 14, 1908, August 23 1909, and January 6 1914, Northern Territory Archives, F302; personal communication, Vern O'Brien; records at the Office of the Placenames Committee, Darwin.

18 N. Pearce, 1978: 30.

19 'Death of Mr. H.C. Trew', *Adelaide Advertiser*, September 19 1958.

20 Ibid.

21 In *Sands & McDougall's South Australian Directory* (Sands & McDougall Pty. Ltd., 1884-1936), Trew is listed in 1932 as a 'station manager, Olary', but in following years no occupation was listed and his address was given as '21 Pier St., New Glenelg.'

22 R.M. Williams referred to 'Compie' Trew as 'Pompey Trew'. The name 'Pompey' was also used by Keith Willey in his book, *The Drovers* (The Macmillan Company of Australia Pty. Ltd., Melbourne, 1982: 122), but Willey acknowledges R.M. Williams as one who helped him with the book, so he may have learned the name from him.

23 R.M. Williams was interviewed on my behalf by Jill Bowen, the editor of the *Stockman's Hall of Fame paper*. Jill took notes of this interview and forwarded his remarks to me.

24 K. Willey, 1982: 22.

25 A.M. Duncan Kemp, *Our Sandhill Country,* Angus & Robertson Ltd., Sydney, 1933: 72-73.

26 Obituary for Duncan Kemp's sister, Mrs. Laura Duncan, *The Pastoral Review*, August 16 1955: 1015.

27 A. Marshall, 1937: 21.

28 *Stockman's Hall of Fame paper*, March 1984: 12.

29 Personal communication, Helen Tolcher, April 19 2001, who interviewed Artie Rowland at Nockatunga station in August 1975 and asked him about the origin of broncoing. Rowland died in Brisbane in 1984 at the age of 98 (*Stockman's Hall of Fame paper*, March 1984: 12).

30 'Death of Mr. H.C. Trew', *Adelaide Advertiser,* September 19 1958.

31 H.G. Lamond, November 18 1963: 1187.

32 Trew's article did not use the term 'bronco branding'. It is clear that Lamond remembered reading about Trew's 'lasso system' and applied the name by which Trew's modified system later became known.

Chapter 5

1 H.C. Trew, 'Trew's Lasso System', *The Pastoralists' Review*, March 15 1905: 18.

2 Ibid.

3 In a letter Thorold Grant wrote to *The Pastoralists' Review* after Trew's article appeared, he mentioned that Trew had gone from Glen Helen to Clifton Hills ('Trew's Lasso System', *The Pastoralists' Review,* April 15 1905: 130-31). Trew himself wrote two letters to *The Pastoralists' Review* in response to a correspondent who criticised his 'lasso system', and he gave his address as 'Goyder's Lagoon', a station north of and adjacent to Clifton Hills ('Trew's Lasso System', *The Pastoralists' Review*, July 15 1905: 388 and 'Trew's Lasso System', November 15 1905: 726). According to S.E. Pearson, Clifton Hills was sometimes referred to as Goyder's Lagoon ('Through the Pleistocene', *The Pastoral Review*, January 16 1928: 53).

4 Charlie Rayment said that he had improvised exactly the same system using a couple of gidgee trees.

5 H.C. Trew, *The Pastoralists' Review*, March 15 1905: 21.

6 W. Thorold Grant, April 15 1905: 130-31. Grant states in his letter that at the time of Trew's experiment Glen Helen station was owned by Grant and Stokes, these were respectively his father and uncle (*The Graziers' Review*, October 16 1925: 887).

7 E. W. Parke and C.H. Walker took up Henbury in October 1876. At some stage Parke's brother became a partner. Walker died in 1885 and drought forced the Parke brothers off the place in the mid to late 1890s (*Northern Territory Trust News*, vol. 6, no. 1, February 1989: 8-9).

8 J. Cummings, 'Trew's Lasso System', *The Pastoralists' Review*, May 15 1905: 209.

9 H.C. Trew, July 15 1905: 388.

10 J. Cummings, 'The Lasso System', *The Pastoralists' Review*, September 15 1905: 578.

11 W. Thorold Grant, 'The Lasso System', *The Pastoralists' Review*, November 15 1905: 726.

12 H.C. Trew, November 15 1905: 726.

Chapter 6

1 G.H. Lamond, *Tales of the Overland*, Hesperian Press, Perth, 1986: 41.

2 'Wild West Carnival', *Adelaide Advertiser,* December 22 1890.

3 First published in serialised form in *Athenaeum* magazine, beginning March 1 1865, published in book form in 1866 by Chapman, Hall and Bentley, London.

4 B. Buchanan, *In the Tracks of Old Bluey*, Central Queensland University Press, Rockhampton, 1997: 3.

5 J. Mora, 1987: 62-63.

6 J. Monaghan, *Australians and the Gold Rush: California and Down Under, 1849-1854,* University of California Press, Berkley, 1966: 240.

7 J. Molony, *Eureka*, Melbourne University Press, Melbourne, 2001: 94 (first published by Viking, 1984).

8 C.D. Ferguson, 'Eye-witness report from a California Ranger', in N. Keesing (ed), *Gold Fever*, Angus and Robertson, Sydney, 1967: 229.

9 Obituary for Austin Mack, in *The Pastoral Review*, July 15 1918: 639.

10 Ibid.

11 'Warrego', 'Australians in Mexico', *The Pastoral Review*, June 11 1940: 578.

12 'Personal' (obituary), *The Graziers' Review*, October 16 1923: 770 .

13 W. Lees, *Coaching in Australia: A History of the Coaching Firm of Cobb & Co.*, The Carter Watson Co., Brisbane, 1917: 13, 20; C. Fetherstonhaugh, 1917: 60-61.

14 T. Hall, *The Early History of Warwick District and Pioneers of the Darling Downs,* Vintage Books, Toowoomba, 1988: 33.

15 Anon, 'A Reminiscence of 1863. – North Queensland', *The Pastoralists' Review,* December 15 1905. Whoever it was who sent this item to the *Pastoralists' Review* made reference to 'Mr Drew's one man lassoing, and hauling a calf up to posts', clearly a reference to Compie Trew's original article.

16 'The World's Circus', *Sydney Morning Herald*, December 17 1888; 'World's Circus' (advertisement), *Sydney Morning Herald*, December 15 1888.

17 (Advertisement) *Sydney Morning Herald*, March 22 1890; 'Texas Jack's First Performance', *Sydney Morning Herald*, April 7 1890.

18 *Sydney Morning Herald*, March 24 1890.

19 C. Barrett and H. Vallance, 'The wild west show: socio-historic spectacle and characters as circus', *Australasian Drama Studies*, No. 35, October 1999: 120-128.

20 W. Thorold Grant, April 15 1905.

21 'Dr. W. F. Carver's Wild America', *The Argus*, December 20 1890.

22 'Dr. W. F. Carver's Wild America', *The Argus*, January 15 1891.

23 'Wild America', *The Argus*, December 24 1890.

24 Copy of the program for Wirth's Circus in 1890, supplied by Mr F.W. Braid, Ballina.

25 P. Wirth, *The Life of Philip Wirth: A Lifetime with an Australian Circus*, Troedel and Cooper, Melbourne, 1935: 49-56.

26 M. St Leon, *Spangles and Sawdust: the Circus in Australia*, Greenhouse Publications, Pty. Ltd., Richmond (Victoria), 1983.

27 'Entertainments. Musical and Dramatic', *Adelaide Observer*, December 27 1890: 26.

28 *Adelaide Observer*, January 2 1892.

29 G. Manning, *Manning's Place Names of South Australia*, privately published, 1970: 216.

30 H.C. Trew, January 15 1906: 895.

31 Ibid; Biographical notes on Chapman, supplied by Rans Baker, researcher at the Carbon County Museum, Wyoming.

32 Biographical notes on Chapman.

33 H.A. Chapman, 'Cattle Ranching in the United States', *The Pastoralists' Review*, December 15 1905: 798-799.

34 'Ranch Life in Wyoming, U.S.A.', *The Pastoralists' Review*, July 16 1906: 390

Chapter 7

1 Information on the meaning of the word 'bronco' was supplied by the Australian National Dictionary Centre at the Australian National University, and this shows that a well-broken horse could be called a 'bronco horse' from at least 1869.

2 Personal communication from Julia Robinson at the Australian National Dictionary Centre, Canberra, who checked various American dictionaries on my behalf. However, a search on the internet has revealed that the word 'broncoing' has recently come into use in American English. One meaning concerns a technique used in BMX bike riding which involves bouncing or rocking the bike as though it is bucking. The other meaning concerns the bucking motion of a person engaged in sexual intercourse.

3 R. Percival, 'Cattle Raising in the Western States' [of the USA], *The Australasian Pastoralists' Review*, May 15 1895: 168 (reprinted from the *Canterbury Times*).

4 O'Grady, 'Lasso Making', *The Pastoralists' Review*, August 15 1907: 514-15.

5 'On the Wallaby. Warenda Station, Queensland', *The Pastoralists' Review*, July 15, 1909: 454.

6 H.K.V. Hungerford, *The Australian Stockman*, June 1914: 160. Mitchell Library, C857. This is a handwritten manuscript in which Hungerford describes his experiences on Innamincka station in 1913.

7 D.C. Thompson to the Administrator of the Northern Territory, nd, Australian Archives CRS A3 N.T. 1914/2576.

8 '"Broncoing" in East Kimberley, Western Australia', *The Pastoralists' Review*, April 14 1927: 319.

9 Personal communication from Julia Robinson at the Australian National Dictionary Centre, Canberra.

10 Marion Beattie, *Wild Dust and Green Fields: Bronco George Hits the Trail*, Research Publications Pty. Ltd., Melbourne, 1991. Marion is a granddaughter of Bronco George.

11 Research notes provided by Marion Beattie.

12 *The Morning Bulletin* (Rockhampton), October 22 1895; *The Northern Miner* (Charters Towers), June 27 1895 and June 28 1895.

13 Advertisement in *The Croydon Golden Age*, December 21 1896.

14 'Among the greatest of horsemen', *The Daily Mercury* (McKay, Qld.), October 12 1983: 11.

15 'Bronco George, the World Famed Circus Proprietor. A True American Yarn: Bronco George's Experiences', *The Showman*, August 6 1909. *The Showman* was the official organ of travelling showmen and entertainers in 1909.

16 'Broncho George's Roughriders', *The Barrier Truth* (Broken Hill), May 18 1906; 'Broncho George's Roughriders', *Adelaide Advertiser*, March 28 1906.

17 'Broncho George – Champion Horseman Interviewed', *Hoofs and Horns*, June 1946: 18-20.

18 Sam Rolleston, cited in an unprovenanced news cutting from a McKay paper.

19 Personal communication, Marion Beattie.

20 Ibid.

21 *The Daily Mercury* (Mackay, Qld.), October 12 1983: 11.

22 Ibid.

23 R. Edwards, letter to the editor of the *Stockman's Hall of Fame paper*, June 1986: 12. Along with this letter is an illustration of 'Humphrey's Breaking in Saddle' which Edwards says was a 1919 model.

24 Advertisement in *The Graziers' Review*, October 16 1922: 637.

25 R.S. Muir, *Pituri Pete: A Saga of Cattle Lands and Desert Sands*, self published, 1987: 150.

26 T. Jordan, 1993: 210-211.

Chapter 8

1 H.K.V. Hungerford, 1914: 160.

2 Personal communication, Phil Gee, who has seen an example in the Lake Eyre country.

3 Personal communication, Norm Forster, Killarney station, Victoria River district1999.

4 Personal communication, Norm Forster, Timber Creek races, 1999.

5 Personal communication, Norm Forster, Killarney station, Victoria River district, 1999.

6 Personal communication, Norm Forster, Timber Creek races, 1999.

7 Personal communication, Lester Cain, at his property Swanvale in western Queensland, June 11 2000.

8 For example, see A. M. Ingham, *Pioneers of the Kimberleys: The Maggie Lily Story*, Halstead Press, Sydney, 2000: 94.

9 Personal communication, Charlie Rayment, at Eildon Park station, western Queensland, June 12 2000.

10 Personal communication, Norm Forster, Timber Creek races, 1999.

11 For example, see B. Simpson, *Packhorse Drover*, ABC Books, Sydney, 1996, photograph on page 70.

12 Personal communication, Charlie Rayment, Eildon Park station, June 12 2000.

13 Personal communication, Buck Buchester, recorded at Kalkarindji, Northern Territory, in 2000. Buck has spent more than fifty years working cattle in the Victoria River district and is widely recognised as an expert 'bronco man'.

14 J.K. Little, ('Culkah')'The Cattleman's Work: Ropes and Branding', *The Pastoral Review*, September 16 1942: 615. Little was on Victoria River Downs when Jack Watson was manager, in 1895-96. Some sources pronounce the name of this ratchet device as 'Robinson's' and others pronounce it as 'Robertsons'.

15 Personal communication, Charlie Rayment, Eildon Park station, June 12 2000.

16 'Q55', August 16 1955: 1006.

17 H.G. Lamond, 'Branding Methods', *The Pastoral Review*, February 24 1964: 109.

18 Personal communication, Charlie Schultz; L.A. Miller, *The Border and Beyond: Camooweal 1884-1984*, Privately published, 1984: 80.

Chapter 9

1 H.C. Trew, 'Cattle Ranching in America', *The Pastoralists' Review*, January 15 1906: 895; 'Death of Mr. H.C. Trew', *Adelaide Advertiser*, September 19 1958.

2 *The Pastoralists' Review*, July 15, 1909: 454.

3 A. Marshall, 1937: 21.

4 H.K.V. Hungerford, 1914: 160.

5 'New Method of Branding Calves', *The Morning Bulletin* (Rockhampton) April 28 1906.

6 D.C. Thompson, Australian Archives CRS A3 N.T. 1914/2576.

7 Australian Investment Agency Collection, Charles Darwin University, Darwin, Accession No. 91292. S.C. 1767, 'Musters camp and Broncho Branding. Nero Yards Aug 1923'.

8 Australian Investment Agency Collection, Charles Darwin University, Darwin, Accession No. 91292. S.C. 1886, 'Broncho Branding - Bradshaws Run 1923'.

9 'Australian Stockmen Yarding Cattle', *Town & Country Journal*, March 6 1875; H.W. Haygarth, 1848: 59; B. Simpson, 1996: 17.

10 Personal communication, Jan Cruickshank, Bob Watson's granddaughter.

11 Ibid; see also *The Graziers' Review*, February 16 1925: 1451 and June 16 1925: 381.

12 *Northern Territory Times*, June 22 1900.

13 Jack Watson to Goldsbrough Mort & Co. Ltd., December 5 1895. Goldsbrough Mort and Co. Ltd., Report on VRD. Goldsbrough Mort & Co: Board Papers, 1893-1927. Noel Butlin Archives, Australian National University, 2/124/1659.

14 Dyson Lacy, 'Reports on several Queensland stations', 1903, Stockman's Hall of Fame collection, Longreach, HF 473.

15 Personal communication, Jack Sammon, April 2005.

16 Private correspondence, Charlie Rayment, May 20 2002.

17 Chris Hughes, *In the first Hundred Years*, unpublished history of Nockatunga station, 1972.

18 *The Pastoralists' Review*, July 15, 1909: 454.

19 A. Marshall, 1937: 21.

20 Personal communication, Charlie Rayment, Eildon Park station, June 12 2000.

21 J.K. Little ('Culkah'), 'Unknown York Peninsula', part 2, *The Pastoral Review*, October 16 1953: 1069.

22 J.K. Little ('Culkah'), 'The Life of a Cattleman', part 6, 'To The Northern Market', *The Pastoral Review*, November 16 1932: 1068.

23 N.A. McLennan,. *Ord River Station, W.A. Yesterday and Today*. Privately Published, Victoria, 1965: 4-5 (Letter from John McLennan, manager of Ord River station, to his bother Malcolm in Victoria, April 16 1896).

24 Jack Watson, December 5 1895. Noel Butlin Archives, Australian National University, 2/124/1659.

25 R. Watson to Goldsbrough Mort and Co. Ltd., March 1899. Goldsbrough Mort and Co. Ltd: Reports on station properties, 1898-1901. Noel Butlin Archives, Australian National University, 2/307/1.

26 J. Gunn, *We of the Never Never*, Hutchinson, London, 1907: 144-45.

27 'Magenta Joe', 'Some Pastoral Notes and Comments from an Old Stockman's Point of View', *Northern Territory Times*, October 27 1905. Magenta Joe' was John Dunn who died of fever in the Palmerston hospital in December 1909 (*Northern Territory Times*, December 13 1907).

28 J. Makin, *The Big Run*, Weldon Publishing, Sydney, 1992: 90-91; *Northern Territory Times*, June 22 1900.

29 T. Ronan, 1962; *Northern Territory Times*, June 22 1900.

30 'A Wet Trip – Splendid Rains – A Sudden Death', *Northern Territory Times,* February 5 1904.

31 T. Ronan, 1962: 145; A. Marshall, 1937: 21.

32 'H7H' (Hely Hutchinson), 'Odd Stock and Other Notes', *The Morning Bulletin* (Rockhampton), June 20 1905.

33 'Station Life', *Northern Territory Times,* June 10 1910.

34 W.L. McDonald, unpublished manuscript in the possession of his daughter, Mrs. Jill Campbell, of Kybo station, on the Nullabor Plain in Western Australia.

35 T. Ronan, 1962: 218.

36 T. Ronan, 1964: 13.

37 Personal communication, Laurie Bain, April 2005.

38 M. Durack, *Kings in Grass Castles*, Corgi, Sydney, 1986: 285-86; M. Durack, *Sons in the Saddle*, Corgi, Sydney, 1985: 433-34.

39 'Q55', August 16 1955: 1006.

40 H. Pearce, *Lakefield National Park Cultural Heritage Report, Volume 2, Lakefield-Laura Station Historical Heritage Places Survey*, Cultural Heritage Branch, Queensland Department of Environment and Heritage, 1998, sections on Couch Grass Yard and Rocky Yard.

41 I was told about these yards and provided with a sketch of one by Philip Gee, an expert camel man with a deep knowledge of the Lake Eyre country.

42 H.G. Lamond, 'Branding Methods', *The Pastoral Review*, March 18 1964: 213.

43 A.S. Wiseman, 'The Botterill [sic] Branding Frame', *The Pastoral Review*, October 16 1942: 699.

44 'A Calf-branding Machine', *The Pastoralists' Review*, February 15 1907: 1026.

45 J.B. Sharp, 'The Cattleman's Working Outfit. Use of the Botterill Frame', *The Pastoral Review*, September 16 1942: 621.

Chapter 10

1 J.H. Kelly, 1949(?): 51-54.

2 'Portable Branding Yards', *The Pastoralists' Review*, November 16 1908: 799.

3 Personal communication, Jack Sammon.

4 J. Perkins and J. Thompson, June 1992: 122.

5 H.G. Lamond, 'Branding Methods', *The Pastoral Review*, January 17 1964: 19

6 *Hergott Herald*, vol. 5, no. 5, July 1989: 13.

7 Personal communication, R.M. Williams, whose company began manufacture of this innovation.

8 Anon, 'The Expert: Branding Calves by the Bronco Method', *Hoofs and Horns*, November 1964: 4.

9 Personal communication, Lester Cain, Swanvale station, western Queensland, June 11 2000.

10 Anon, ('The Expert'), November 1964: 4.

11 Charlie Rayment, recorded at Eildon Park station, between Winton and Jundah, June 12 2000.

12 R. Laxalt and O. Mazzatenta, 'Last of a Breed: The Gauchos', *National Geographic*, October 1980: 486; N.R. Hughes, 'Corriento Harness: The Gaucho Saddles Up', *The Pastoral Review*, March 22 1965: 242-43.

13 Ibid: 242.

14 C. Roderick, *Leichhardt the Dauntless Explorer,* Angus and Robertson Publishers, Sydney, 1988: 439.

15 H.G. Lamond, *Horns & Hooves: Handling Stock in Australia*, Country Life Ltd., London, 1931: 174.

16 H.G. Lamond, *The Pastoral Review*, November 18 1963: 1187.

17 Personal communication, Helen Tolcher, April 19 2001, citing part of an interview she conducted with Artie Rowland at Nockatunga station in August 1975, about the origin of broncoing.

18 'O'Grady, August 15 1907: 514.

19 B. Grant, *Encyclopedia [sic] of Rawhide and Leather Braiding*, Cornell Maritime Press, Inc. Cambridge (USA), 1972: 30-33.

20 Apart from the article by 'O'Grady, some of the other descriptions of greenhide rope-making are (anon) 'Lasso Making', *The Stockowner's Guide*, The Pastoralists' Review Proprietary Limited, Sydney, 1912: 225-27; A.N.M., 'Preparing the Hide: Rope Making', *The Pastoral Review*, March 16 1934: 263; Jack Knox, 'Making Rawhide Ropes: Advice from Northern Australia', *West Australian* (undated cutting, but post 1936); J.K. Little ('Culkah'), 'The Cattleman's Work: Ropes and Branding Pens', *The Pastoral Review*, September 16 1942: 615-16; 'The Xpert: Branding Calves by the Bronco Method', *Hoofs and Horns*, November 1944: 4; J.K. Little 'Culkah', 'Greenhide Gear and How to Make It', *The Pastoral Review*, part 1, December 15 1951: 1368-69 and part 2, January 18 1952: 21-22; 'The Xpert Says: Rawhide Ropes', *Hoofs and Horns*, March 1955: 2; 'Q55', 'Operations at Calf Branding', *The Pastoral Review*, July 16 1955: 871-73; R.M. Williams, *The Bushman's Handcrafts*, R.M. Williams Pty Ltd, Adelaide, 1977 (first published 1943): 90-94; and R. Edwards, *Australian Traditional Bush Crafts*, Lansdowne Press, Melbourne, 1975: 55-56.

21 The Victoria River Downs records at the Noel Butlin Archives, Australian National University, place Knox on VRD from October 1933 to September 1938 (Victoria River Downs Ledgers, 1909-1944, Ledger 4, February 1930 to February 1937, Goldsbrough Mort Papers, 42/15).

22 C. Rayment, *Greenhide Rope Making*, unpublished typescript in possession of the author.

23 R.M. Williams, 1977: 90.

24 J.K. Little, December 15 1951: 1368-69; 'Q55', July 16 1955: 872.

25 R.M. Williams, 1977: 90.

26 Personal communication, Mick Bower and Buck Buchester, Katherine 1999.

27 R.M. Williams, 1977: 90.

28 'Lasso Making', *The Stockowner's Guide*, The Pastoralists' Review, Sydney, 1912: 225.

29 Personal communication, Mick Bower, Katherine, 1999.

30 'O'Grady', August 15 1907: 515.

31 J.K. Little, January 16 1952: 21.

32 Personal communication, Charlie Rayment.

33 J.K. Little, January 16 1952: 21.

34 Personal communication, former drover and ringer Rodney Watson.

35 Personal communication, Ernie Rayner, Atherton tableland, February 2005. Ernie went to the Victoria River district in the late 1950s and worked as a ringer on Coolibah, Victoria River Downs and Willeroo. Later he was a Stock Inspector in the district.

36 'Q55', July 16 1955: 871-73; J.K. Little ('Culkah'), September 16 1942: 615; personal communication, Rodney Watson.

37 'Nowan', 'Who's Who in the Bushland: Acacia Species (The Gidgees)', *The Pastoral Review*, July 15 1939: 770-71.

38 R. Macnamara, *The Way it Was*. Privately published, Qld, 2002: 26.

39 Personal communication, Mick Bower and Buck Buchester, Katherine 1999.

40 J.K. Little, September 16 1942: 616.

41 A leather plough, also known as a leather gauge, is an adjustable tool that enables a strip of uniform width to be cut from a sheet of greenhide or leather.

42 R. Edwards, 1975.

43 Personal communication, Ernie Rayner, February 2005.

44 Charlie Rayment, 'Greenhide Rope Making', unpublished typescript, August 8 1994.

45 J.K. Little, January 16 1952: 21.

46 *The Stockowner's Guide*, 1912: 225.

47 'O'Grady, 1907: 515.

48 Personal communication, Charlie Rayment.

49 R.M. Williams, 1977: 93.

50 J. Knox, *The West Australian*, nd.

51 J.K. Little, January 18 1952: 21-22.

52 Personal communication, Mick Bower, Katherine, February 1999.

53 R. Edwards, 1975: 55-56.

54 R.M. Williams, 1977: 92.

55 Personal communications, Charlie Rayment and Mick Bower.

56 'The Xpert Says: Rawhide Ropes', *Hoofs and Horns*, March 1955: 2; R. Edwards, 1975: 56.

57 'O'Grady', August 1907: 515.

58 Personal communication, Mick Bower, Katherine, February 2005. Mick went to work on Birrindudu station in 1954 and was a ringer on Gordon Downs, Nicholson, Nutwood Downs and Manbulloo. While on Birrindudu he was taught to make greenhide ropes by Jack Burns, a man who's skill in making greenhide ropes was recognised throughout the region.

59 R. Macnamara, 2002: 26; Personal communication, Mick Bower, Katherine, 1999.

60 'Q55', July 16 1955: 871-873.

61 R. Macnamara, 2002: 26.

62 Personal communication, Charlie Schultz, owner of Humbert River station from 1927 to 1941. Charlie Rayment also drags his ropes along a dirt road behind a vehicle to remove the hair.

63 J.K. Little, December 15 1951: 1369;

64 A.N.M., 1934: 263; R. M. Williams, 1977: 93; R. Edwards, 1975: 56.

65 R. Macnamara, 2002: 26.

66 'O'Grady', August 15 1907: 515.

67 'Q55', July 16 1955: 871; Personal communication, Charlie Rayment.

68 Personal communication, Buck Buchester, Katherine 1999.

69 In a report written by VRD manager Alf Martin in 1929, he reported that 25,000 calves had been branded the previous year. The number branded rose to 33,000 in 1933 (Noel Butlin Archives, Bovril Australian Estates Ltd. Records, 119/4/1, Correspondence between Bovril Australian Estates Ltd., London, and station manager, 1927-33).

70 Personal communication, Rob Sampson, a head stockman on Wave Hill in the 1950s.

71 Personal communications, Buck Buchester, Charlie Rayment and Rob Sampson.

72 Personal communication, Mick Bower, Katherine, February 2000.

Chapter 11

1 G.O. Walker, 'Upholding a Tradition', *Outback*, issue 14, December 2000/January 2001: 146.

2 H.G. Lamond, December 18 1963: 1304.

3 Personal communication, Lloyd Fogarty, Timber Creek. Lloyd has lived in the Victoria River district for more than fifty years and was manager of Auvergne station from 1955 to 1979.

4 G. McLaren, *Big Mobs: The Story of Australian Cattlemen*, Fremantle Arts Centre Press, Fremantle, 2000: 133. McLaren notes that this statement may be something of an exaggeration as bronco horses were generally eased into the work.

5 Charlie Rayment, recorded at Eildon Park station, between Winton and Jundah, June 12 2000.

6 Private correspondence, Charlie Rayment's comments on the first draft of this book, March 7 2005.

7 Lester Cain, recorded at his property Swanvale, between Winton and Jundah, June 11 2000.

8 N. 'Pic' Willetts with Merice Briffa, *Wind on the Cattle: Recollections of Fifty Years Droving*. Auscribe Enterprises, Oxley, 2002: 51.

9 Buck was recorded at Kalkarindji, Victoria River district, Northern Territory, in 2000. He arrived in the district in about 1952 and has worked there ever since, achieving a reputation as a very good horseman and bronco man.

10 'Three To Go', *Mack Down Under*, (the Mack Truck Bulldog Bulletin), 1919-1994 75th Anniversary: 23.

11 Lester Cain, recorded at his property Swanvale, June 11 2000.

12 Ibid.

13 S.E. Pearson, 'The Land of Lots of Time', *The Pastoral Review*, October 16 1926: 916.

14 Personal communication, Laurie Bain, April 2005.

15 Laurie Bain from South Tibraden, Western Australia, interviewed by telephone in April 2005.

16 Charlie Rayment, recorded at Eildon Park station, June 12 2000

17 Lester Cain, recorded at his property Swanvale, June 11 2000.

Chapter 12

1 N. 'Pic' Willetts with Merice Briffa, 2002: 17.

2 S. Raymond, 1977: 37.

3 N.A. McLennan, 1965: 4-5.

4 M. McKenzie, letter to the Editor, *The Pastoral Review*, June 16 1932: 559.

5 J.K. Little ('Culkah'), 'The Cattleman's Working Outfit', *The Pastoral Review*, July 16 1942: 578.

6 H.C. Trew, March 15 1905: 20.

7 H.C. Trew, January 15 1906: 895.

8 *The Morning Bulletin* (Rockhampton), April 28 1906.

9 'H7H' (Hely Hutchinson), 'A Big Cattle Trip: From Western Australia to Queensland', *The Morning Bulletin* (Rockhampton), August 29 1905.

10 Jim White, March 2002.

11 Len Carrington, 1977.

12 A. Marshall, 1937: 21.

13 Lester Cain, recorded at his property Swanvale, June 11 2000

14 Ted Fogarty interviewed by Lindsay Rice, on 'The Country Hour', ABC Radio, Alice Springs, 1998 (the Fogarty family was still broncoing on Lila Creek, Palmer Valley and Lucy Creek stations when this interview was recorded).

15 Buck Buchester, recorded at Kalkarindji, Northern Territory, 2000.

16 Ibid; Buck's story was confirmed by Gavin Hoad (personal communication), manager of Wave Hill station, who got the story direct from Peter Sherwin.

Chapter 13

1 W. Cowan, *'Rollin' Yer Swag'*, Boolarong Publications, Brisbane, 1985: 127.

2 Lester Cain, recorded at his property Swanvale, June 11 2000.

3 Ibid.

4 L. Simmons and D. Lewis, *Kajirri, the Bush Missus*, Central Queensland University Press, Rockhampton, 2005.

5 J. Illingsworth, (ed) *A Christmas Card in April*, James Cook University, Townsville, 1990: 121-22. This book contains the memoirs of Eric and Thelma Martel and the event described here occurred on Strathleven station, Cape York, in 1956.

6 Charlie Rayment, recorded at Eildon Park station, June 12 2000.

7 Lester Caine, recorded at his property Swanvale, June 11 2000.

8 Charlie Rayment, recorded at Eildon Park station, June 12 2000.

9 R.S. Muir, 1987: 152, 155.

Chapter 14

1 Personal communications, Ted Fogarty, Darryl Hill, Joyce Galvin, Lindsay Ward and Nick Gill; R. Chaney, 'Broncoing Lives On', *Outback*, February-March, 2001: 72-74.

2 Personal communication, Andrew Glenn, manager of Ashburton station; see *Outback Magazine*, issue 15, February-March 2001: 73.

3 Personal communication, Alan Fennel, manager of Lambina station, June 2005.

4 Personal communication, Ted Fogarty, Alice Springs.

5 Personal communication, Rodney Watson, who made inquiries at Rosebud on my behalf.

6 Information from notes supplied by Nick Gill who interviewed Phil Stoker, co-owner of Marion Downs, in 1997.

7 Personal communication, Lindsay Ward, Halls Creek, 2000.

8 BTEC stands for the Brucellosis and Tuberculosis Eradication Campaign.

9 P.J. Schmidt, *Beef Cattle Production*, Butterworths, Sydney, 1985: 260-61.

10 R.S. Muir, 1987: 151.

11 'The Territory opts for bronco branding', *Stockman's Hall of Fame paper,* September 1983: 4; 'Bronco Branding Challenge From N.T.', *Stockman's Hall of Fame paper*, March 1984: 2.

12 J. Bowen, 'Bronco Branding Blast Off!', *Stockman's Hall of Fame paper*, June 1984.

13 Personal communication, Des Stenhouse, who was one of the contestants.

14 Personal communication, Randall Crozier, manager of Anna Creek station, northern South Australia.

15 'A Tribute to the Duracks', *Stockman's Hall of Fame paper*, December 2001.

BIBLIOGRAPHY

Allen, G. *Stockman's Hall of Fame paper*, December 1997.

'A.N.M.', 'Preparing the Hide: Rope Making', *The Pastoral Review*, March 16 1934.

Anning, H. 'Cattle Branding Yards', *The Pastoral Review*, April 16 1931.

Anon, 'A Reminiscence of 1863. – North Queensland', *The Pastoralists' Review*, December 15 1905.

Anon, 'From Most Humble Beginnings: How the First beef cattle came to Australia', *The Australian Shorthorn*, January 1956.

Anon, 'Lasso Making', *The Stockowner's Guide*, The Pastoralists' Review Proprietary Limited, Sydney, 1912.

Anon, 'Three to Go', *Mack Down Under* (the Mack Truck Bulletin) 1919-1994, 75th Anniversary: 23.

Anon, 'On the Wallaby. Warenda Station, Queensland', *The Pastoralists' Review*, July 15 1909: 454.

Anon. *The Romance of the Stockman*, Penguin Books Australia Ltd., Sydney, 1993.

Barrett, C. and H. Vallance, 'The wild west show: socio-historic spectacle and characters as circus', *Australasian Drama Studies*, No. 35, October 1999.

Beattie, M. *Dry Dust and Green Fields: Bronco George Hits The Trail,* Research Publications Pty, 1991.

Bonser, K.J. *The Drovers*, Macmillan and Co. Ltd., London, 1970.

Bowen, J. 'Bronco Branding Blast Off!', *Stockman's Hall of Fame paper*, June 1984.

Buchanan, B. *In the Tracks of Old Bluey*, Central Queensland University Press, Rockhampton, 1997.

Campbell, A. 'Men and Horses of Today', *The Pastoral Review*, June 16 1951.

Cannon, M. *Life in the Country: Australia in the Victorian Age: 2*, Thomas Nelson, Melbourne, 1973.

Carrington, L. 'Reminiscences of Augustus Downs During 1920s & 30s', in P. Carrington, *How Many Grids to Gregory?,* Gregory Branch of the CWA, 1977.

Carter, J. *In the Tracks of the Cattle*, Angus and Robertson, Sydney, 1968.

Caswell, A. *Hints from the Journal of an Australian Squatter*, Smith, Elder & Co., London, 1843.

Chaney, R. 'Broncoing Lives On', *Outback*, February-March, 2001.

Chapman, H. 'Cattle Ranching in the United States', *The Pastoralists' Review*, December 15 1905.

Chapman, H. 'Ranch Life in Wyoming, U.S.A.', *The Pastoralists' Review*, July 16 1906.

Colyer, R.J. *The Welsh Cattle Drovers*, University of Wales Press, Cardiff, 1976.

Cowan, W. *'Rollin' Yer Swag'*, Boolarong Publications, Brisbane, 1985.

'Culkah' (J.K. Little), 'The Life of a Cattleman', part 6, 'To The Northern Market', *The Pastoral Review*, November 16 1932.

'Culkah', 'The Cattleman's Working Outfit', *The Pastoral Review*, July 16 1942.

'Culkah', 'The Cattleman's Work: Ropes and Branding Pens', *The Pastoral Review*, September 16 1942.

'Culkah', 'Our Future Cattlemen', *Pastoral Review*, April 17 1950: 348-49.

'Culkah', 'Greenhide Gear and How to Make It', *The Pastoral Review*, part 1, December 15 1951 and part 2, January 18 1952.

'Culkah', 'Greenhide Gear and How to Make It', *The Pastoral Review*, part 2, January 18 1952.

'Culkah', 'Unknown York Peninsula', part 2, *The Pastoral Review*, October 16 1953.

Cummings, J. 'Trew's Lasso System', *The Pastoralists' Review*, May 15 1905.

Cummings, J. 'The Lasso System', *The Pastoralists' Review*, September 15 1905.

Cunningham, P. *Two Years in New South Wales*, Henry Colburn, London, 1827.

Delbridge, A. *et al* (eds), *The Macquarie Dictionary*, The Macquarie Library Pty. Ltd., second edition, 1991.

de Satge, O. *Pages from the Journal of a Queensland Squatter*, Hurst and Blackett Ltd., London, 1901.

Duncan Kemp, A.M. *Our Sandhill Country,* Angus & Robertson Ltd., Sydney, 1933.

Durack, M. *Sons in the Saddle*, Corgi Books, Sydney, 1985.

Durack, M. *Kings in Grass Castles*, Corgi Books, Sydney, 1986.

Edwards, R. *Australian Traditional Bush Crafts*, Lansdowne Press, Melbourne, 1975.

Edwards, R. *Stockman's Hall of Fame paper*, June 1986.

Fairholme, George Knight Erskine, 'Branding Calves. Australia [No. 11]', *Fifteen Views of Australia in 1845, by G.K.E.F.*, comprising fifteen hand-coloured lithographs, Mitchell Library.

Ferguson, C.D. 'Eye-witness report from a California Ranger', in N. Keesing (ed), *Gold Fever*, Angus and Robertson, Sydney, 1967.

Fetherstonhaugh, C. *After Many Days: Being the Reminiscences of Cuthbert Fetherstonhaugh*, E.W. Cole, Melbourne, 1917.

Finch-Halton, H. *Advance Australia!*, W. H. Allen & Co., London, 1885.

Fletcher, B. *Landed Enterprise and Penal Society: A History of Farming and Grazing in New South Wales Before 1821*, Sydney University Press, Sydney, 1976.

Grant, B. *Encyclopedia of Rawhide and Leather Braiding*, Cornell Maritime Press, Inc., Cambridge (USA), 1972.

Grant, W.T. 'The Lasso System', *The Pastoralists' Review*, November 15 1905.

Grant, W.T. 'Trew's Lasso System', *The Pastoralists' Review*, April 15 1905.

Gunn, J. *We of the Never Never*, Hutchinson, London, 1907.

Haldane, A.R.B. *The Drove Roads of Scotland*, Thomas Nelson and Sons Ltd., London, 1952.

Hall, T. *The Early History of Warwick District and Pioneers of the Darling Downs*, Vintage Books, Toowoomba, 1988.

Harris, A. *Settlers and Convicts*, Melbourne University Press, Melbourne, 1953.

Haygarth, H.W. *Recollections of Bush Life in Australia During a Residence of Eight Years in the Interior*, John Murray, London, 1848.

Hendersen, J. *Excursions and Adventures in New South Wales*, Saunders and Otley, London, 1854.

Howard, K. *Stockman's Hall of Fame paper*, March 2001.

Hughes, C. *In the first Hundred Years*, unpublished history of Nockatunga station, 1972.

Hughes, N.R. 'Corriento Harness: The Gaucho Saddles Up', *The Pastoral Review*, March 22 1965.

Hungerford, H.K.V. 'The Australian Stockmen', Mitchell Library, hand written manuscript, C857, 1914.

'H7H' (Hely Hutchinson), 'Odd Stock and Other Notes', *The Morning Bulletin* (Rockhampton), June 20 1905.

'H7H' (Hely Hutchinson), 'A Big Cattle Trip: From Western Australia to Queensland', *The Morning Bulletin* (Rockhampton), August 29 1905.

Illingsworth, J. (ed) *A Christmas Card in April*, James Cook University, Townsville, 1990.

Ingham, A.M. *Pioneers of the Kimberleys: The Maggie Lily Story*, Halstead Press, Sydney, 2000.

Jordan, T. *North American Cattle-Ranching Frontiers: Origins, Diffusion, and Differentiation*, University of New Mexico Press, Albuquerque, 1993.

Kelly, J.H. *Northern Australia Cattle Survey*, Bureau of Agricultural Economics, Department of Commerce and Agriculture, Canberra, 1949.

Knox, J. 'Making Rawhide Ropes: Advice from Northern Australia', *West Australian* (undated cutting, but post 1936).

Lacy, D. 'Reports on several Queensland stations', 1903, Stockman's Hall of Fame collection, Longreach, HF 473.

Lamond, H.G. *Horns & Hooves: Handling Stock in Australia*, Country Life Ltd., London, 1931.

Lamond, H.G. 'Branding Methods', *The Pastoral Review*, November 18 1963.

Lamond, H.G. 'Branding Methods', *The Pastoral Review*, December 18 1963: 1304.

Lamond, H.G. 'Branding Methods', *The Pastoral Review*, January 17 1964: 19-20.

Lamond, H.G. 'Branding Methods', *The Pastoral Review*, February 24 1964: 109.

Lamond, H.G. 'Branding Methods', *The Pastoral Review*, March 18 1964.

Lamond, G.H. *Tales of the Overland*, Hesperian Press, Perth, 1986.

Laxalt, R. and Mazzatenta, O. 'Last of a Breed: The Gauchos', *National Geographic*, October 1980.

Lees, W. *Coaching in Australia: A History of the Coaching Firm of Cobb & Co.*, The Carter Watson Co., Brisbane, 1917.

McDonald, W.L., unpublished manuscript in the possession of his daughter, Mrs. Jill Campbell, of Kybo station, on the Nullarbor Plain in Western Australia.

McKenzie, M. (letter to the editor), *The Pastoral Review*, June 16 1932: 559.

McLaren, G. *Big Mobs: The Story of Australian Cattlemen*, Fremantle Arts Centre Press, Fremantle, 2000.

McLennan, N.A. *Ord River Station, W.A. Yesterday and Today*. Privately Published, Victoria, 1965.

McWhiney, G. and McDonald, F. 'Celtic Origins of Southern Herding Practices', *The Journal of Southern History*, vol. 51, no. 2, May 1985.

Macnamara, R. *The Way it Was*. Privately published, Queensland, 2002.

'Magenta Joe', 'Some Pastoral Notes and Comments from an Old Stockman's Point of View', *Northern Territory Times*, October 27 1905.

Makin, J. *The Big Run*, Weldon Publishing, Sydney, 1992.

Manning, G. *Manning's Place Names of South Australia*, privately published, 1970.

Marshall, A. 'Bronchoing!', *Walkabout*, February 1st, 1937.

Miller, L.A. *The Border and Beyond: Camooweal 1884-1984*, Privately published, 1984.

Molony, J. *Eureka*, Melbourne University Press, Melbourne, 2001 (first published by Viking, 1984).

Monaghan, J. *Australians and the Gold Rush: California and Down Under, 1849-1854*, University of California Press, Berkley, 1966.

Mora, J. *Trail Dust and Saddle Leather*, Bison Books, 1987: 131-32 (originally published by University of Nebraska Press, Lincoln, 1946).

Morey, E. 'Timber Creek Patrols, Pt. 5', *Northern Territory Newsletter*, 1978.

Muir, R.S. *Pituri Pete: A Saga of Cattle Lands and Desert Sands*, self published, 1987.

Mundy, C. *Our Antipodes*, Richard Bentley, London, 1852.

'Nowan', 'Who's Who in the Bushland: Acacia Species (The Gidgees)', *The Pastoral Review*, July 15 1939.

'O'Grady, 'Lasso Making', *The Pastoralists' Review*, August 15 1907.

O'Loughlen. F. 'Australia Needs More Men Like Jack Kilfoyle'. *The Australian Shorthorn*, August 1952.

Pearce, H. *Homesteads of the Stony Desert*, Rigby, Adelaide, 1978.

Pearce, H. *Lakefield National Park Cultural Heritage Report, Volume 2, Lakefield-Laura Station Historical Heritage Places Survey*, Cultural Heritage Branch, Department of Environment and Heritage, 1998.

Pearson, S.E. 'The Land of Lots of Time', *The Pastoral Review*. October 16 1926.

Pearson, S.E. 'Through the Pleistocene', *The Pastoral Review*, January 16, 1928.

Perkins, J. and Thompson, J. 'The Technology of Open-Range Cattle Farming in Early European Australia', *Prometheus*, vol. 10, no. 1, June 1992.

Percival, R. 'Cattle Raising in the Western States' [of the USA], *The Australasian Pastoralists' Review*, May 15 1895 (reprinted from the *Canterbury Times*).

'Q55', 'Operations at Calf Branding', *The Pastoral Review*, July 16 1955: 871-73.

'Q55', 'Operations at Calf Branding', *The Pastoral Review*, August 16 1955: 1006.

Rayment, C. 'Greenhide Rope Making', unpublished typescript, August 8th 1994 (copy in possession of the author).

Raymond, S. (ed), *Tourist to the Antipodes: 'William Archer's Australian Journey, 1876-77'*, University of Queensland Press, Brisbane, 1977.

Reid, T.M. *The Headless Horseman*, Chapman, Hall and Bentley, London, 1866.

Roderick, C. *Leichhardt the Dauntless Explorer,* Angus and Robertson Publishers, Sydney, 1988.

Ronan, T. *The Deep of the Sky*, Cassell & Co. Ltd., Sydney, 1962.

Ronan, T. *Packhorse and Pearling Boat*, Cassell & Co. Ltd., Sydney, 1964.

Sands & McDougall's South Australian Directory, Sands & McDougall Pty. Ltd., 1884-1936, 1942-1973.

Schmidt, P.J. *Beef Cattle Production*, Butterworths, Sydney, 1985.

Shann, E.O.G., *Cattle Chosen*, Oxford University Press, London, 1926.

Sharp, J.B. 'The Cattleman's Working Outfit. Use of the Botterill Frame', *The Pastoral Review*, September 16 1942.

Sidney, S. *The Three Colonies of Australia*, Ingram, Cooke and Co., London, 1852.

Simmons, L. and D. Lewis *Kajirri, the Bush Missus*, Central Queensland University Press, Rockhampton, 2005.

Simpson, B. *Packhorse Drover*, ABC Books, Sydney, 1996.

St Leon, M. *Spangles and Sawdust: the Circus in Australia,* Greenhouse Publications, Pty. Ltd., Richmond (Victoria), 1983.

Terry, M. *Through a Land of Promise: with gun, car and camera in the heart of Northern Australia*, Herbert Jenkins Ltd., London, 1927.

'The Xpert: Branding Calves by the Bronco Method', *Hoofs and Horns*, November 1944.

'The Xpert Says: Rawhide Ropes', *Hoofs and Horns*, March 1955.

Thompson, J. and Perkins, J. 'The Wild Cowpastures Revisited', *Journal of the Royal Australian Historical Society*, vol. 77, pt. 4, June 1992.

Tolcher, T. *Innamincka: The Town with Two Lives*, Innamincka Progress Association, 1990.

Trew, H.C. 'Trew's Lasso system', *The Pastoralists' Review*, March 15 1905.

Trew, H.C. 'Trew's Lasso System', *The Pastoralists' Review*, July 15 1905.

Trew, H.C. 'Trew's Lasso System', *The Pastoralists' Review*, November 15 1905.

Trew, H.C. 'Cattle Ranching in America', *The Pastoralists' Review*, January 15 1906.

The Macquarie Illustrated World Atlas, The Macquarie Library Pty. Ltd., Sydney, 1984.

Walker, G.O. 'Upholding a tradition', *Outback*, issue 14, December 2000-January 2001.

'Warrego', 'Australians in Mexico', *The Pastoral Review*, June 11 1940.

Watts, C. *Stockman's Hall of Fame paper*, June 1999.

Wentworth, W.C. *Statistical, Historical, and Political Description of the Colony of New South Wales*, G. and W. Whittaker, London, 1819.

White, J. 'Brunette Downs 1928-45', *Stockman's Hall of Fame paper*, March 2002.

White, H.F. 'Branding Performances', *The Pastoral Review*, June 16 1932.

Willetts, P. with M. Briffa, *Wind on the Cattle: Recollections of 50 Years Droving*, privately published, 2002.

Willey, K. *The Drovers,* The Macmillan Company of Australia Pty. Ltd., Adelaide, 1982.

Williams, R.M. *The Bushman's Handcrafts*, Griffin Press Limited, Adelaide, 1978 (first published 1943).

Wirth, P. *The Life of Philip Wirth: A Lifetime with an Australian Circus*, Troedel and Cooper, Melbourne, 1935.

Wiseman, A.S. 'The Botterill Branding Frame', *The Pastoral Review*, October 16 1942.

Wright, P.P. *72 Years in Australia and the South Pacific*, 1919. Mitchell Library manuscript 1452 or microfilm copy CY 1902.

Website

Australian Bureau of Statistics – Australia's beef cattle industry.

Private Photograph Collections

Gerry Ash Collection
Laurie Bain Collection
Marion Beattie Collection
Mick Bower Collection
Bobbie Buchanan Collection
Lester Caine Collection
John Graham family Collection
Mildred Iverach (nee Martin) Collection
Ray Macnamara Collection
Marie Mahood Collection
Howard Pearce Collection
Charlie Schultz family Collection
Maureen Wood Collection
Bob Young Collection

Public Photograph Collections

Bovril Collection, Edinburgh City Archives, Scotland.
F.H. Johnston Collection, Australian National Library.
Oxley Library, Brisbane.

R.T. Maurice Collection, Royal Geographical Society of South Australia, Adelaide.
Stockman's Hall of Fame, Longreach, Qld.
Australian Investment Agency Collection, Charles Darwin University, Darwin.
Townshend Collection, Mitchell Library, Sydney.

Radio Interviews

Ted Fogarty interviewed by Lindsay Rice, on 'The Country Hour', ABC Radio, Alice Springs, 1998.

Newspapers and Journals

The Adelaide Advertiser, December 1890, March 1906, September 1958.
The Adelaide Chronicle, August 1939.
The Adelaide Observer, December 1890, January 1891, January 1892.
The Argus, December 1890; January 1891.
The Australian Shorthorn, January 1956, October, 1950 or 1959.
The Barrier Truth (Broken Hill), May 1906.
The Croydon Golden Age, December 1896.
The Daily Mercury (McKay, Qld), October 1983.
The Graziers' Review, October 1922, October 1923, February 1925, June 1925, October 1925.
The Hergott Herald, July 1989.
Hoofs and Horns, November 1944, June 1946, November 1957.
The Morning Bulletin (Rockhampton), October 1895, April 1906.
The Northern Miner (Charters Towers), June 1895.
The Northern Territory Times, June 1900 and February 1904, December 1907, June 1910.
Mack Down Under, (the Mack Truck Bulldog Bulletin), 1919-1994, 75th Anniversary.
Northern Territory Trust News, February 1989.
Outback Magazine, February-March 2001.
The Pastoral Review, July 1905, November 1905, December 1905, July 1906, February 1907, November 1908, July 1909, July 1918, January 1923, April 1927, June 1940, November 1953, August 1955.
The Australasian Pastoralists' Review, May 1892.
The Showman, August 1909.
The South Australian Register, January 1866.
The Stockman's Hall of Fame paper, September 1983, March 1984, December 2001.
The Sydney Morning Herald, December 1888; March 1890, April 1890.
The Town & Country Journal, March 1875.

National Archives of Australian

D.C. Thompson to the Administrator of the Northern Territory, CRS A3 N.T. 1914/2576.
Noel Butlin Archives, Australian National University.
Jack Watson to Goldsbrough Mort & Co. Ltd., December 5 1895. Goldsbrough Mort and Co. Ltd., Report on VRD. Goldsbrough Mort & Co: Board Papers, 1893-1927, 2/124/1659.
R. Watson to Goldsbrough Mort and Co. Ltd., March 1899. Goldsbrough Mort and Co. Ltd: Reports on station properties, 1898-1901, 2/307/1.
Victoria River Downs Ledgers, 1909-1944, Ledger 4, February 1930 to February 1937, Goldsbrough Mort Papers, 42/15.
Bovril Australian Estates Let. Records. Correspondence between Bovril Australian Estates Ltd., London, and station manager, 1927-33, 119/4/1.

Northern Territory Archives

Timber Creek Police Journal, 1908, 1909, and 1914, Northern Territory Archives, F302.

INDEX

A

B

C

D

E

F

G

H

I

Books by

Darrell Lewis

Roping in the History of Broncoing

The Murranji Track